Electronic Structure of Materials

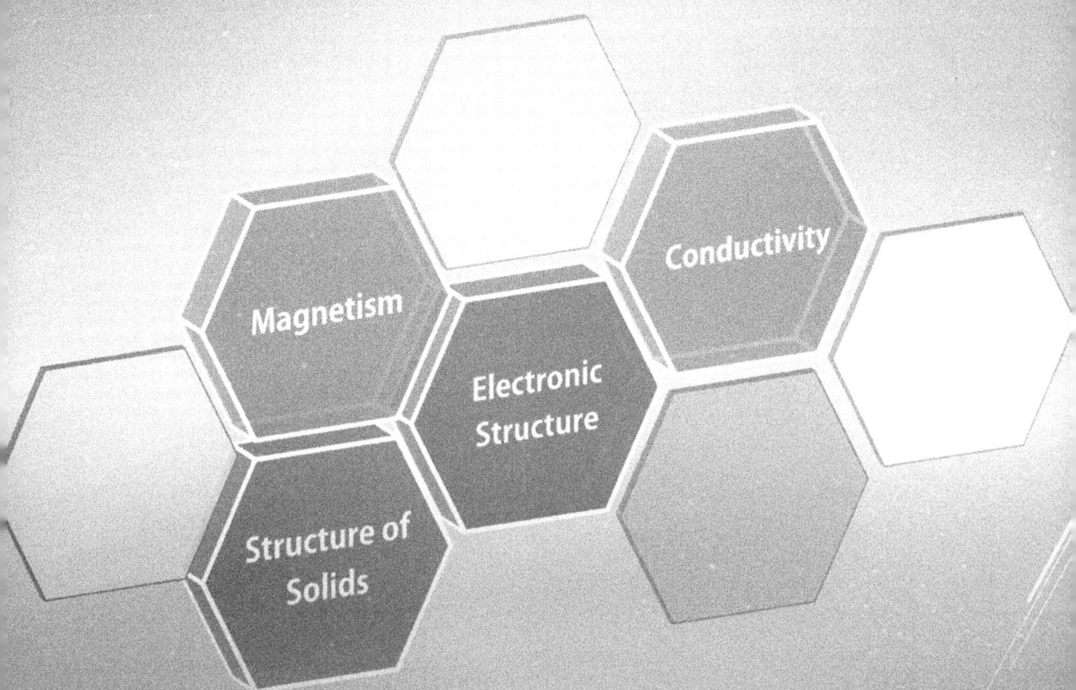

Electronic Structure of Materials

Challenges and Developments

edited by

Natalia V. Chezhina
Dmitry A. Korolev

PAN STANFORD PUBLISHING

Published by

Pan Stanford Publishing Pte. Ltd.
Penthouse Level, Suntec Tower 3
8 Temasek Boulevard
Singapore 038988

Email: editorial@panstanford.com
Web: www.panstanford.com

British Library Cataloguing-in-Publication Data
A catalogue record for this book is available from the British Library.

Electronic Structure of Materials: Challenges and Developments

ISBN 978-981-4800-55-6 (Hardcover)
ISBN 978-0-429-24287-8 (eBook)

Contents

Preface

This book shows the specialists in the field of materials chemistry the importance and necessity of studying the electron structure of solid oxide systems, promising from the point of view of application in various fields. The everlasting problem of selecting the qualitative and quantitative composition of materials may be solved only on the basis of the exhaustive knowledge of their crystal and electron structure and of the model simulation of the influence of various elements on the target properties of oxide ceramics. A great diversity of crystal structures and physical and chemical properties of oxide systems used as materials nowadays made the authors to concentrate on a certain number of systems that are the most popular in modern material technologies. All the parts of the book are associated with the method of magnetic dilution as the most universal method for studying the electron structure of solids. The method was advanced as early as in 1960, in particular, by Professor S. M. Ariya and is still being developed in Saint Petersburg state University by the group of magnetochemistry of solids. In these years, the mathematical and physical apparatus of the magnetic dilution method has been developed together with the development of the theory of magnetochemistry. The book shows the expediency of using various physical methods with the aim of revealing the chemical structure of particular solid oxide systems depending on their chemical composition and crystal structure and provides a clue for selecting optimal compositions for obtaining special properties of target materials.

The authors express their profound gratitude to Evgenii Shchipunov, whose highest skills were provided for all the experiments, especially the operation of magnetic devices. We are grateful to Professors V. N. Pak, R. S. Bubnova, S. K. Filatov, and S. I Andronenko, Associate Professors M. G. Krzhizhanovskaya, N. V. Platonova, and I. A. Kasatkin, Dr. S. V. Nekipelov (Syktyvkar State University), and scientific coworker D. A. Spiridonova for scientific cooperation at various stages of studies. Special thanks to Professor D. G. Kellerman (Institute of Solid State Chemistry, RAS, Ekaterinburg) for magnetization studies. Our special acknowledgement goes to the coworkers of Research Park of St. Petersburg University: Center for X-Ray Diffraction Studies and Center for Chemical Analysis and Material Research.

Natalia V. Chezhina
Dmitry A. Korolev

Introduction

Nowadays oxide materials have acquired great interest among researchers and in the industry. Gone are the times when ceramics were associated with building materials, utensils, and other even less prestigious applications. Now complex oxides are widely used as materials for oxygen sensors and oxygen-conducting membranes of catalytic reactors, and photocatalyzers in the UV and visual regions of the spectrum. Recent discoveries of high-temperature superconductivity in oxide systems of colossal magneto resistance in lanthanum manganites gave a great impetus to the search of new compounds demonstrating unusual and promising physicochemical properties. The challenges in the field of energy sources started a completely new and wide problem of development in the compositions for the technologies of solid oxide fuel cells (SOFC).

All the above-mentioned oxide systems have various structures, and their applications are different. Moreover, almost all the systems are multicomponent, which means that they are solid solutions including transition and rare earth elements somehow distributed over various sublattices in the complex oxides. A large series of structures, especially perovskite like, are tolerant to substitution varying both by the nature of the substituent and by their concentration. Up to now the search for optimal compositions has been carried out purely experimentally, which resulted in a wide series of published works recommending better compositions for a particular property, e.g. electron-ionic conductivity, ionic conductivity, magneto resistivity, and so on. Taking into account the fact that in the Periodic Table of elements the number of possible dopants in the metal sublattice amounts to about a hundred, and sometimes several elements are introduced simultaneously, the more so, the ratio between several dopants may vary in wide regions, the number of possible compositions tends to infinity.

Therefore, the problem of selecting the compositions of oxide ceramics requires attaining theoretical grounds. A large number of questions must be answered in doing this. In the systems containing *d*- or *f*-elements, these questions may be summed up as following if not exclusively.

1. What are the oxidation and spin states of paramagnetic elements, and how does the crystal structure of the oxide system influence them and how does the electron structure of paramagnetic elements influence the crystal structure of oxide materials?
2. What are the interatomic interactions between paramagnetic atoms, and what effect do they have on the physical and chemical properties of oxide systems?
3. How do diamagnetic elements of the oxide system influence both above-mentioned characteristics?
4. How do all these factors affect the performance (magnetic, electric, catalytic, and so on) of the target materials?

In other words, with the aim of laying the pathway to selecting the composition of ceramics with predetermined physical and chemical characteristics, we must solve the problems thoroughly connected with each other in various ways:

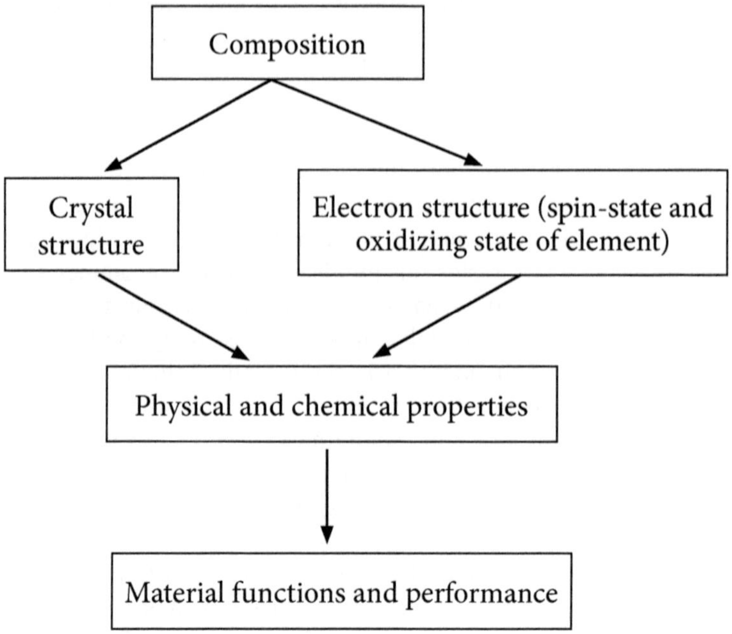

The interrelation between the composition (both qualitative and quantitative) and the crystal structure is successfully studied by the X-ray method exquisitely developed nowadays and allowing the structure refinement to be made for powder samples. The problem of electron structure and its interrelation with crystal structure and physicochemical properties is more complicated and can be solved by a number of physical methods sensitive to the state of paramagnetic atoms and interatomic interactions. All of them have special features when applied to solid state and various limits of their application.

For example, such an extremely sensitive physical method as electron spin resonance (ESR) could be used for solving the above-mentioned problems, but not all the paramagnetic centers give signals in ESR. Ni^{2+} is active in ESR, but not always, mostly when the atom is in a regular octahedral surrounding, Ni^{3+} low spin give a good signal, but Ni^{3+} high spin does not. Mn^{2+} and Mn^{4+} are active in ESR even at room temperature, but Mn^{3+} may be seen only at very low temperatures, close to liquid helium. This means that we can use this method only if we have a signal in ESR, but if we have no signal we cannot conclude sometimes that there is no paramagnetic atom in the particular oxidation or spin state. Nuclear magnetic resonance has even narrower application, since only a limited number of atoms have nonzero nuclear spin and the presence of paramagnetic atoms in the system extremely complicates the resolution of the spectra and their interpretation. Mossbauer spectra can be successfully applied to iron-containing systems, no other paramagnetic elements can be detected. However, even with iron the method has some limitations associated with other diamagnetic elements, such as lead and strontium, which damp down the signal of iron.

In other words, all these physical methods, sensitive to the electron structure in themselves, may be used only as auxiliary methods for studying particular oxide systems, the more so that there always must be some regularities in electron structures and properties as function of the origin of paramagnetic element e.g. of its position in the Periodic table. The same can be applied to optic spectra in the IR or visual region widely used for the interpretation of the chemical structure of complex compounds, since in the solids owing to exchange line breadth the interpretation of the results may be dubious.

The method of static magnetic susceptibility may be reckoned as a universal method for studying the electron structure of inorganic compounds for the following reasons. Magnetic susceptibility is the value showing the difference between the magnetic fields in vacuum and in a substance. Since this value cannot be equal to zero, the problem consists of adequate analysis of the observed effect and of creating quantum-mechanical models for its interpretation. Let us make a short insight into the fundamentals of magneto chemistry and consider its application in the solid state in more details.

Chapter 1

Fundamentals of Magnetochemistry

Natalia V. Chezhina and Dmitry A. Korolev

Department of General and Inorganic Chemistry, St. Petersburg State University, Universitetskaya nab. 7/9, St. Petersburg, 199034, Russia

natalia.chezhina@gmail.com, n.chezhina@spbu.ru

1.1 Method of Static Magnetic Susceptibility in Coordination Chemistry

Nowadays the theoretical fundamentals of the method of static magnetic susceptibility for studying the electron structure of coordination or complex compounds of *d*- and *f*-elements are developed in substantial details.[1-4] As R. L. Carlin states in his textbook on magnetochemistry[3]: "*The Nobel Laureate J. H. Van Vleck has written a brief but comprehensive review of magnetism [1] [reference 5 in this chapter], and he has argued that quantum mechanics is the key to understanding magnetism. I would go further and argue that magnetism is the key to understanding quantum mechanics, at least in the pedagogical sense, for so many physical phenomena can be understood quantitatively in this discipline.*"

This method allows a large number of parameters of chemical structure to be determined. The most important of them are the

Electronic Structure of Materials: Challenges and Developments
Edited by Natalia V. Chezhina and Dmitry A. Korolev

ISBN 978-981-4800-55-6 (Hardcover), 978-0-429-24287-8 (eBook)
www.panstanford.com

valence and spin state of transition elements in a compound, quantitative characterization of spin transitions.[6] The Hamiltonian describing the behavior of an atom in the magnetic field includes a number of parameters, which can be determined on the basis of examination of temperature dependencies of magnetic susceptibility and effective magnetic moment, needless to say that the number of variables must be small, and other parameters are known from independent experimental data—optic and electron spectroscopy, other magnetic methods. Such parameters are first of all the coordination number of a transition element atom, the energy of crystal field, spin–orbit coupling. Magnetic susceptibility is affected by the distortion of coordination polyhedra and the degree of the bond covalence.

One of the most important applications of magnetic susceptibility method appeared as a response to cluster compound chemistry. It started from the detection of diamagnetism in a number of clusters containing *d*-elements in a low oxidation state and in the course of time developed into a sufficiently accurate theory of exchange interactions in clusters,[2,5] the model of exchange channels,[7] which opens a wide possibility of describing the interatomic interactions in the coordination chemistry. In other words, nowadays the coordination chemistry has such a universal and powerful instrument for studying the chemical (and electron) structure as magnetochemistry. Before trying to adjust it to the study of electron structure of modern materials, which are usually solids, let us consider the fundamentals of magnetochemistry.

Magnetic phenomena observed upon introducing a substance into magnetic field are *paramagnetism* and *diamagnetism*. Paramagnetism is the property associated with unpaired electrons, and diamagnetism is associated with filled electron orbitals. This means that on measuring any paramagnetic compound we always observe a summarized effect including both a paramagnetic and a diamagnetic contributions. With the aim of simulating the magnetic behavior of paramagnetic atoms, we must separate the paramagnetic effect in the pure state.

Diamagnetism is the effect of closed electron shells, which are present also in paramagnetic atoms. The magnetic field results in the closed type currents their magnetic moments being directed in the opposite direction to the magnetic field. Diamagnetic susceptibility is determined by the number, shape, size, and orientation of electron orbitals and does not depend on the field strength and on temperature. For example, for monatomic gases equation (Eq. 1.1) is obeyed:

$$\chi_A = -N\frac{e^2}{6\pi mc^2}\sum r^2, \quad (1.1)$$

where e is the electron charge, m electron mass c the velocity of light, and r the average orbital radius.

Paramagnetism associated with unpaired electrons does not depend on magnetic field but essentially depends on temperature. In the case of ideal paramagnetic, the paramagnetic component of magnetic susceptibility obeys Curie law (Eq. 1.2) or Curie–Weiss law (Eq. 1.3):

$$\chi_M = C/T, \quad (1.2)$$

$$\chi_M = C/(T-\theta), \quad (1.3)$$

where χ_M is the paramagnetic component of magnetic susceptibility calculated per 1 mole of paramagnetic atom (emu·mol^{-1}), C the Curie constant (K·emu·mol^{-1}), and θ the Weiss constant (K·emu·mol^{-1}).

In the case of the systems containing paramagnetic elements, it is essential that as a rule the paramagnetic effect exceeds the diamagnetic effect by several orders of magnitude. Thus the paramagnetic component of magnetic susceptibility of iron(III) containing compounds calculated per mole of Fe^{3+} at 100 K is 44000 × 10^{-6} emu·mol^{-1}, whereas its diamagnetic correction is only −10 × 10^{-6} emu·mol^{-1}. This allows the paramagnetic component of magnetic susceptibility to be calculated by subtracting the sum of diamagnetic corrections from molar susceptibility of the substance:

$$\chi_M = M_{\chi g} - \sum_i \chi_i^{dia}, \quad (1.4)$$

where χ_g is the measured specific magnetic susceptibility (emu·g^{-1}), M the molar mass of the compound under study (g·mol^{-1}), and $\sum_i \chi_i^{dia}$ the sum of diamagnetic corrections for all the atoms in the compound (emu · mol^{-1}).

The most informative value for a chemist is the so-called effective magnetic moment μ_{eff}, which is connected to the paramagnetic component of magnetic susceptibility by Eq. (1.5):

$$\chi_{\text{M}} = \frac{N_{\text{A}}\mu_{\text{eff}}^2\beta^2}{3kT}, \tag{1.5}$$

where N_{A} is Avogadro's number (6.02 × 10^{-23} mol^{-1}), β Bohr magneton (0.9273 × 10^{-20} erg · G^{-1} = 4.6686 × 10^{-5} cm^{-1} · G^{-1}), and k Boltzmann's constant (1.38044 × 10^{-16} erg · K^{-1} = 0.69504 cm^{-1} · K^{-1}).

Then we obtain

$$\mu_{\text{eff}} = \sqrt{3kT\chi_{\text{M}}/(N_{\text{A}}\beta^2)} \text{ or } \mu_{\text{eff}} = 2.828\sqrt{\chi_{\text{A}}T}. \tag{1.6}$$

μ_{eff} has a comprehensible physical meaning, since it is related by a simple relationship with the atom or ion spin, i.e., with the number of unpaired electrons. The magnetic moments result from two main causes: the electron spins and the orbital movement of electrons. The greatest contribution to the total effective magnetic moment is due to the spin. According to quantum mechanics the so-called "spin only" effective magnetic moment ($\mu_{\text{S.O.}}$) is determined by Eq. (1.7):

$$\mu_{\text{S.O.}} = \sqrt{4S(S+1)} \text{ or } \mu_{\text{S.O.}} = g\sqrt{S(S+1)}, \tag{1.7}$$

where S is the total spin of a paramagnetic atom and g the g-factor or splitting factor, which for a free electron is 2.

Since the total spin S is a multiple to ½ we can obtain the relationship between the magnetic moment and the number of unpaired electrons (n):

$$\mu_{\text{S.O.}} = \sqrt{n(n+2)}, \tag{1.8}$$

Thus it appears rather easy to determine the spin by μ_{eff} and, therefore, the degree of oxidation of a paramagnetic atom. For example, the experiment gave μ_{eff} ~ 3.8 μB in a compound containing manganese atoms, this corresponds to three unpaired electrons, and, consequently, we deal with Mn^{4+}.

The spin only effective magnetic moments for all the possible spins for d-elements are given as follows:

S	1/2	1	3/2	2	5/2
$\mu_{S.O.}$	1.730	2.828	3.873	4.899	5.916

The electron movement round the nucleus results in an orbital moment, which also contributes in the total magnetic moment:

$$\mu_L = \sqrt{L(L+1)},$$

where L is the orbital quantum number of an atom.

In such a case for isolated atoms or ions, the total magnetic moment is equal to

$$\mu_{S+L} = \sqrt{4S(S+1) + L(L+1)}.$$

However, the effective magnetic moments for real complex ions only insignificantly differ from spin only values and are rather far from the values of the total magnetic moment (Table 1.1).[2]

Table 1.1 Effective magnetic moments for the compounds containing 3d-elements

Configuration	Cations	Ground state of free ion	Values of magnetic moment (μB)		
			μ^*_{S+L}	μ^*_S	$\mu^{\#}_{eff}$
d^1	Ti^{3+}, V^{4+}	^2D	3.00	1.73	1.6–1.8
d^2	V^{3+}	^3F	4.47	2.83	2.7–2.9
d^3	V^{2+}, Cr^{3+}, Mn^{4+}	^4F	5.50	3.87	3.7–3.9
d^4	Cr^{2+}, Mn^{4+}	^5D	5.58	4.90	4.8–5.0
d^5	Mn^{2+}, Fe^{3+}	^6S	5.92	5.92	5.9–6.0
d^6	Fe^{2+}	^5D	5.48	4.90	5.0–5.5
d^7	Co^{2+}	^4F	5.20	3.87	4.4–5.2
d^8	Ni^{2+}	^3F	4.47	2.83	2.9–3.4
d^9	Cu^{2+}	^2D	3.00	1.73	1.8–2.2

*Calculated results.

Experimental data obtained at temperature 300 K.

In the crystal substances the orbital moment is partially or completely quenched owing to the fact that the electric field resulting from crystal surrounding confines the orbital movement of electrons. Then the contribution of orbital component into the effective magnetic moment is introduced as a certain correction determined by the spin–orbit constant (λ) with

$$\lambda = \pm\zeta/(2S),$$

where ζ is the single electron spin orbit coupling constant of a free ion and it's always positive; (+) in this equation relates to half-filled orbitals; (–) is related to the orbitals filled more than a half.

This becomes understandable if we assume that such configurations are "positive holes." ζ for two- and three-charged ions of the elements of the first transition series are given in Table 1.2. These values evidently increase from left to right in this series. For complex compounds we can use the data of Table 1.2 with confidence,[1] but in the solids the contribution of spin–orbit coupling is usually substantially less, it is said that it is "frozen," and for solids the spin–orbit coupling constants must be determined separately.

When studying the compounds containing 3*d*-elements with rather small spin–orbit coupling constants (for 4*d*- and 5*d*-elements ζ are substantially greater than 1000 cm^{-1}) the experimental μ_{eff} appear to be close to the spin only values, for d^1–d^4 configurations being somewhat lower and for d^6–d^9 somewhat higher than $\mu_{S.O.}$ in complete agreement with the sign of λ. A detailed theoretical examination of the effect of spin–orbit coupling leads to the conclusion that the effect of λ is determined by the ground term of the paramagnetic atom.

In the crystal field of octahedral or tetrahedral symmetry, we deal with three ground terms—*A*, *E*, and *T*. *A*—and *E*-terms are orbital singlet and doublet, respectively. These terms are not split by spin–orbit coupling and as a first approximation do not make an orbital contribution. However, if there is a *T*-term with the same multiplicity as *A* and *E* there can be an "admixing" of this term to the ground state. This introduces a certain temperature independent correction (*T*-term is not occupied) to the effective magnetic moment.

Table 1.2 Spin–orbit coupling constants (λ) for the first series of transition elements ions determined on the basis of ζ for free ions

Ion	ζ, cm^{-1}	Number of *d*-electrons	Weak octahedral field		Strong octahedral field		Tetrahedral field	
			Term of ground state	λ (cm^{-1})	Term of ground state	λ (cm^{-1})	Term of ground state	λ (cm^{-1})
Ti^{3+}	155	1	$^2T_{2g}$	+155	$^2T_{2g}$	+155	^2E	+155
V^{3+}	210	2	$^3T_{1g}$	+105	$^3T_{1g}$	+105	$^3A_2$	+105
V^{2+}	170	3	$^4A_{2g}$	+57	$^4A_{2g}$	+57	$^4T_1$	+57
Cr^{3+}	275	3	$^4A_{2g}$	+92	$^4A_{2g}$	+92	$^4T_1$	+92
Cr^{2+}	230	4	^5E_g	+58	$^3T_{1g}$	−115	$^5T_2$	+58
Mn^{3+}	355	4	^5E_g	+89	$^3T_{1g}$	−178	$^5T_2$	+89
Mn^{2+}	300	5	$^6A_{1g}$	—	$^2T_{2g}$	−300	$^6A_1$	—
Fe^{3+}	460	5	$^6A_{1g}$	—	$^2T_{2g}$	−460	$^6A_1$	—
Fe^{2+}	400	6	$^5T_{2g}$	−100	$^1A_{1g}$	—	^5E	−100
Co^{3+}	580	6	$^5T_{2g}$	−145	$^1A_{1g}$	—	^5E	−145
Co^{2+}	515	7	$^4T_{1g}$	−172	^2E_g	−515	$^4A_2$	−172
Ni^{3+}	715	7	$^4T_{1g}$	−238	^2E_g	−715	$^4A_2$	−238
Ni^{2+}	630	8	$^3A_{2g}$	−315	$^3A_{2g}$	−315	$^3T_1$	−315
Cu^{2+}	830	9	^2E_g	−830	^2E_g	−830	$^2T_2$	−830

Term A_1 (d^5 configuration in a weak crystal field and d^6 in a strong field) has no overlying T-term with the same multiplicity. Hence an atom or ion with d^5 configuration must have the spin only effective magnetic moment μ_{eff} = 5.92 μB independent on temperature.

Terms A_2 and E arise from splitting D- and F-terms of the free ion and, consequently, have an overlying T-term with the same multiplicity. For these terms the effect of "admixing" is expressed by the Eq. (1.9):

$$\mu_{eff} = \mu_{S.O.}\left(1 - \alpha\frac{\lambda}{10Dq}\right), \tag{1.9}$$

where α = 2 for E-term, α = 4 for A_2-term, and $10Dq$ is the crystal field splitting in cm^{-1}.

Therefore, we obtain for A_2- and E-terms the effective magnetic moment somewhat differing from the spin only value, but independent on temperature. Then we can calculate g-factor for these ground terms:

$$g = 2\left(1 - \alpha\frac{\lambda}{10Dq}\right) \tag{1.10}$$

The spin–orbit coupling removes the degeneracy of triplet ground states, i.e., T-terms are split by spin–orbit coupling. In such a case in the calculation of effective magnetic moment, we must take into account Boltzmann's occupation of the levels. This inevitably will result in a dependence of μ_{eff} on temperature. In the theoretical calculations μ_{eff} is considered as a function of kT/λ. Theoretical calculations of the effective magnetic moments for T ground terms are described in refs. 1–4. Expressions for theoretical dependencies of μ_{eff} given in Table 1.3.

For approximate estimations of μ_{eff} the variations of effective magnetic moments with temperature for the $^2T_2$, $^3T_1$, $^4T_1$, and $^5T_2$ are given in Fig. 1.1.[1] μ_{eff} for some ions at T = 300 K obtained using λ from Table 1.2 are marked off in the curves. As $kT/\lambda \rightarrow \infty$, i.e., at very high temperatures μ_{eff} approaches μ_{S+L} (see Eq. 1.8).

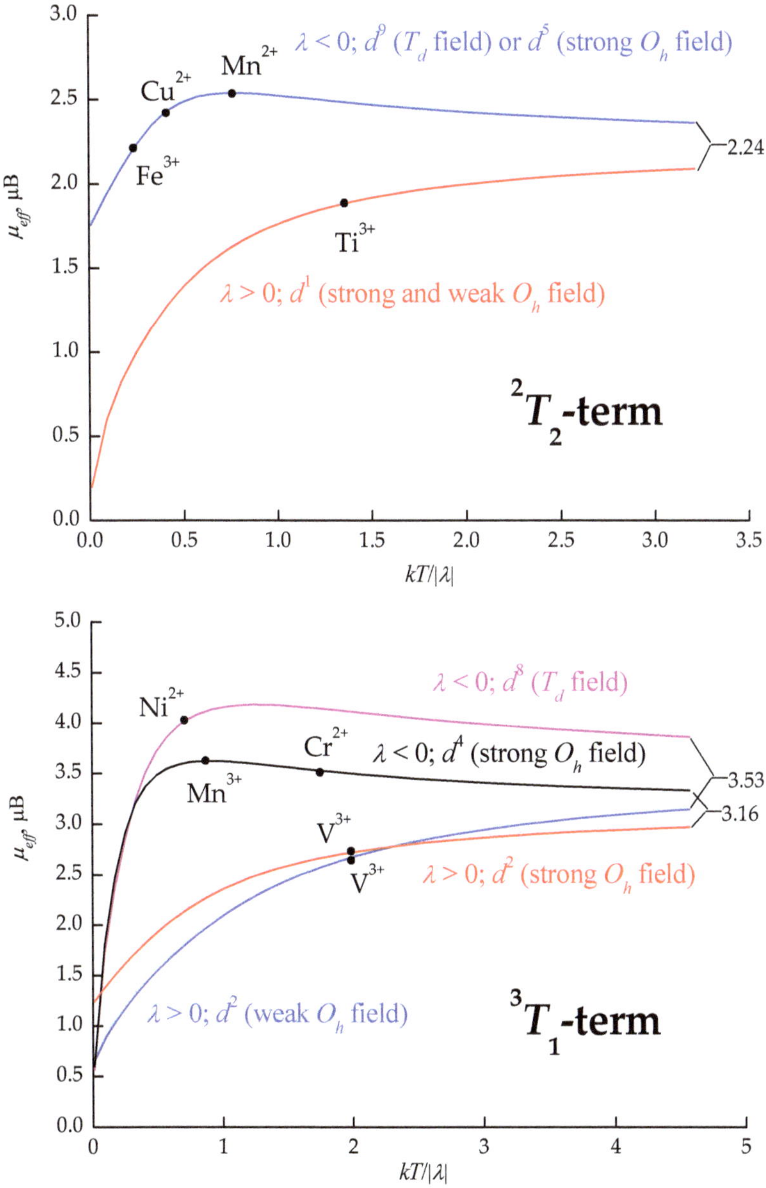

Figure 1.1 Variations of effective magnetic moments with temperature for the triplet terms of ground state in tetrahedral (T_d) and octahedral (O_h) crystal fields.

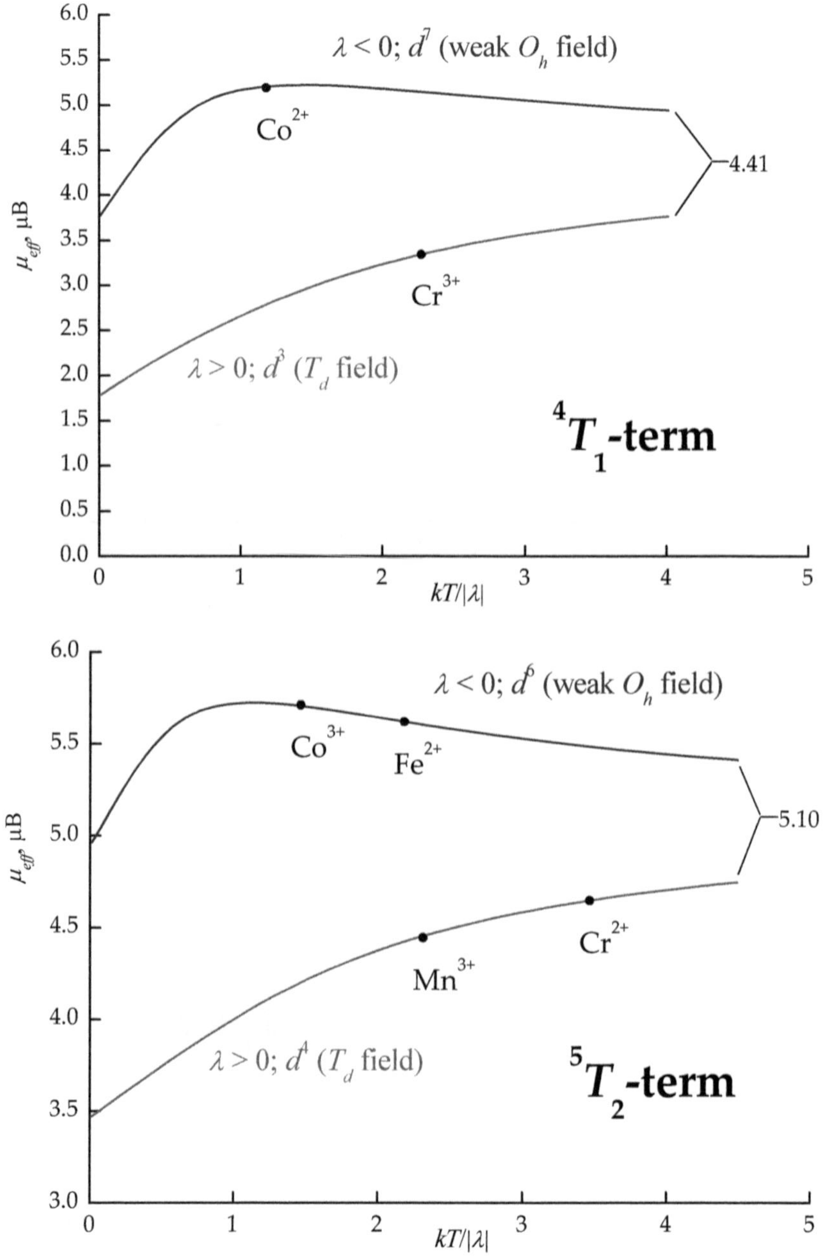

Figure 1.1 (*Continued*)

Table 1.3 Values of effective magnetic moment for *T*-terms and various fields

Term	Expression for μ_{eff}
	Weak field
$^2T_2$	$\mu_{eff}^2 = \dfrac{8+(3y-8)e^{-\frac{3}{2}y}}{y(2+e^{-\frac{3}{2}y})}$
$^3T_1$	$\mu_{eff}^2 = 3\dfrac{0.625y+6.8+(0.125y+4.09)e^{-3y}-10.89e^{-\frac{9}{2}y}}{y(5+3e^{-3y}+e^{-\frac{9}{2}y})}$
$^5T_2$	$\mu_{eff}^2 = 3\dfrac{28y+9.33+(22.5y+4.17)e^{-3y}+(24.5y-13.5)e^{-5y}}{y(7+5e^{-3y}+3e^{-5y})}$
$^4T_1$	$\mu_{eff}^2 = 3\dfrac{3.15y+3.92+(2.84y+2.13)e^{-\frac{15}{4}y}+(4.7y-6.05)e^{-6y}}{y(3+2e^{-\frac{15}{4}y}+e^{-6y})}$
	Strong field
$^2T_2$	$\mu_{eff}^2 = \dfrac{8+(3y-8)e^{-\frac{3}{2}y}}{y(2+e^{-\frac{3}{2}y})}$
$^3T_1$	$\mu_{eff}^2 = 3\dfrac{5y+15+(y+9)e^{-2y}-24e^{-3y}}{2y(5+3e^{-2y}+e^{-3y})}$

Therefore, from the value of μ_{eff} and its dependence on temperature we can determine the degree of oxidation of the central atom, its spin state (high- or low-spin), and the type of coordination polyhedron. For example, in the experiment a certain complex of nickel with coordination number six was obtained, its magnetic susceptibility was measured and the effective magnetic moment was calculated. If μ_{eff} ~ 2.8–3.0 μB and does not depend on temperature, this is Ni^{2+} ($^3A_{2g}$); if μ_{eff} ~ 1.8–2.0 μB and also does not depend on temperature this is low-spin Ni^{3+} (^2E_g); if μ_{eff} > 3.8 μB and increases as the temperature increases, consequently we obtained a high-spin complex of Ni^{3+}.

The coordination number of a transition element atom and the type of coordination polyhedron also can be found by the data of magnetic susceptibility.

The simplest case are the complexes with coordination number four and electron configuration d^8 (Ni^{2+}, Pt^{2+}). If metal atom is in the tetrahedron, its configuration is e^4t^4. It has two unpaired electrons and $\mu_{eff} \sim 2.8–3.0$ μB. However, if the metal atom is in the square planar ligand surrounding, as for example in $[Ni(CN)_4]^{2-}$, such a compound appears to be diamagnetic. However, the situation with the coordination number may appear more complicated, when it is difficult to find how many ligands are coordinated by the transition element atoms.

The matter is that when we are dealing with the octahedron/tetrahedron alternative, the number of unpaired electrons appears to be the same as in the high spin octahedral complexes. Let us try to consider the ground terms of the transition element atoms in the octahedral and tetrahedral surrounding, taking into account the fact that the effective magnetic moment for *A* and *E* ground terms does not depend on temperature, whereas for triplet ground states μ_{eff} is a function of temperature.

As shown in Table 1.4, for all the electron configurations but for d^5 in one of the surroundings (octahedral or tetrahedral) the ground state is not degenerated or doubly degenerated, and in the other it is obligatorily threefold degenerated.

Table 1.4 Ground terms of *d*-element atoms in octahedral and tetrahedral surrounding

d^n	1	2	3	4	5	6	7	8	9
O_h	T_{2g}	T_{1g}	A_{2g}	E_g	A_{1g}	T_{2g}	T_{1g}	A_{2g}	E_g
T_d	E	A_2	T_1	T_2	A_1	E	A_2	T_1	T_2

This means that if for a certain coordination compound sufficiently magnetically diluted to allow the interactions between neighboring paramagnetic atoms to be neglected, a dependence of μ_{eff} on temperature is observed, then either all the paramagnetic atoms or a fraction of them is in the surrounding with triplet ground state. Then using the data of Table 1.3 it is possible to find which surrounding is there. If only a fraction of atoms is in tetrahedral surrounding and the other is octahedral, the experimental values of effective magnetic moment is described by Eq. (1.11):

$$\mu_{\text{eff}}^2 = a\mu_{\text{T}}^2 + (1-a)\mu_{\text{O}}^2, \tag{1.11}$$

where a is the fraction of paramagnetic atoms in the tetrahedral surrounding and μ_{O} and μ_{T} the effective magnetic moments for this very electron configuration in the octahedral and tetrahedral surrounding, respectively.

As mentioned above, by the effective magnetic moment we can find in what state—high-spin (HS) or low-spin (LS)—is the paramagnetic atom. However, in the intermediate crystal fields a situation may arise, when the energies of HS and LS states are close. If the difference between their energies $\Delta E \sim kT$, relative occupations of these states are correlated and depend on temperature. For example, for Co^{3+} (d^6) in a regular octahedral surrounding two states are possible—diamagnetic $^1A_1$ (LS) and paramagnetic $^5T_2$ (HS), the latter term being split by the spin–orbit coupling (Fig. 1.2).

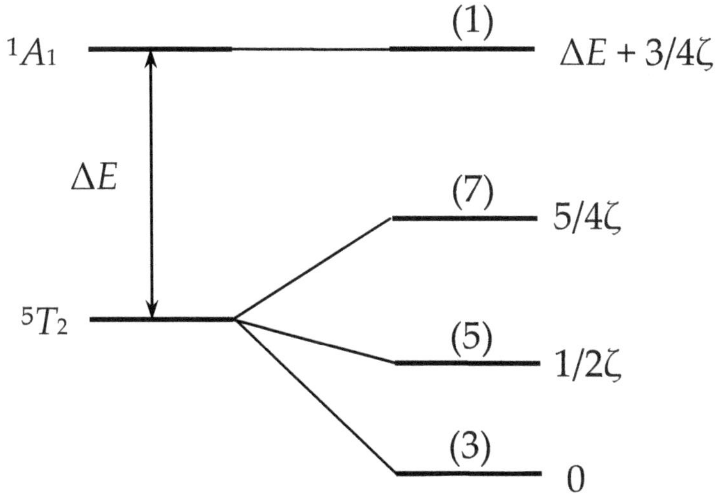

Figure 1.2 Influence of spin–orbit coupling on a splitting of *A*- and *T*-terms.

With respect to Boltzmann's occupation of various states in general, the magnetic susceptibility of such complexes must be described by Eq. (1.12):

$$\chi(T)=\frac{f_1\chi_1(T)+f_2\chi_2(T)\exp(-\Delta E/kT)}{f_1+f_2\exp(-\Delta E/kT)} \tag{1.12}$$

where f_1 and f_2 are the degrees of degeneracy of the states 1 and 2 with the susceptibilities χ_1 and χ_2, respectively; $\Delta E = E_2 - E_1$. In this case it is evident that the sign of ΔE depends on which state is the lowest.

For Co^{3+}, judging from the scheme in Fig. 1.2, ΔE will be positive if the ground state is high-spin and negative if it is low-spin. In the first case, the effective magnetic moment will decrease as the temperature increases, in the second case it will increase.[6] Therefore, by the temperature dependence of the effective magnetic moment the ground state of the atom may be determined with certainty and ΔE may be calculated.

We emphasize that the effective magnetic moment and its dependence on temperature are influenced by such effects as the distortion of coordination polyhedra (an additional splitting of single electron levels) and by the degree of covalency of the metal–ligand bond. These problems are discussed in detail in refs. 2 and 4.

1.2 Exchange Effect and Spin-Spin Interactions

One of the most promising lines of inquiry in modern inorganic chemistry is the chemistry of polynuclear complexes. Such compounds contain interacting paramagnetic ions of transition elements and demonstrate unconventional magnetic properties, which cannot be described as the properties of single atoms. At the same time they are not diamagnetic, which is typical for clusters with metal-metal bonds. Magnetic properties of such compounds are associated with the so-called exchange interactions—a quantum effect resulting in a correlated behavior of the spins of paramagnetic atoms located not far from each other. Magnetic properties of polynuclear complexes are determined by such factors as the electron configuration of a metal, the metal-metal distance, electron structure of ligands, and geometric location of metal and ligands.

For example, binuclear chromium(II) acetate (d^4), $[Cr_2(CH_3COO)_4(H_2O)_2]$ is diamagnetic, whereas a similar complex of copper(II) (d^9) is paramagnetic, though the compounds have the same structure. Similarly, two complexes close in their structure—$[Cr_2Cl_9]^{3-}$ and $[W_2Cl_9]^{3-}$—with the same electron configuration d^3 are quite different in their magnetic properties. The first is paramagnetic, and the second is diamagnetic.

The method of magnetic susceptibility may be an instrument for distinguishing the clusters and describing their electron structure, type, and energy of exchange interactions. The matter is that in the paramagnetic clusters, i.e., clusters, where no complete pairing of the interacting atoms occurs, a certain change in the magnetic behavior must be observed in comparison with single paramagnetic atoms. Thus, the magnetic moment of single Cu^{2+} ion (d^9) does not depend on temperature and is 1.8–2.0 μB. However, if two atoms appear at a small distance from each other and, for example, make up a $[Cu_2(CH_3COO)_4(H_2O)_2]$ cluster, the interaction between them will result in the spins being aligned in parallel to each other or in antiparallel. Two states differing by energy arise. Since the difference in energies of these states is comparable to kT, the effective magnetic moment will depend on temperature.

Two kinds of exchange interactions are recognized: direct and indirect exchange. The direct exchange is the result of overlapping d-orbitals of interacting paramagnetic atoms. The indirect or superexchange occurs via the orbitals of bridge atoms; example, $[(NH_3)_5Cr\text{-}O\text{-}Cr(NH_3)_5]^{4+}$.

In chemistry every interaction is first of all characterized by its energy, in magnetochemistry—also by the sign (ferromagnetic or antiferromagnetic). For quantitative description of interactions in magnetic clusters the Heisenberg–Dirac–Van Vleck model is conventionally used. According to this model Hamiltonian of spin–spin interaction for dimer clusters may be expressed as

$$\hat{H} = -2J_{12}\hat{S}_1\hat{S}_2, \qquad (1.13)$$

where S_1 and S_2 are the angular spin moment of interacting atoms and J_{12} the exchange parameter in cm^{-1}.

If $J_{12} < 0$, the lowest energy level belongs to the state with antiparallel spins ($S = 0$); if $J_{12} > 0$, the lowest energy is for parallel spins. The situation somehow reminds the formation of chemical bond, though it must be remembered that the energy of exchange interactions is much lower than the energy of chemical bond, and only in clusters with metal-metal bond, such as for example $[W_2Cl_9]^{3-}$, the energy of exchange interactions attains large values. Such clusters are usually diamagnetic ($S = 0$).

The exchange parameter is connected with the energy of exchange interactions $E(S')$ by Eq. (1.14):

$$E(S') = -J_{12}[S'(S' + 1) - S_1(S_1 + 1) - S_2(S_2 + 1)], \tag{1.14}$$

where S' is the total spin of the cluster, $S' = (S_1 + S_2), (S_1 + S_2 - 1), \ldots, |S_1 - S_2|$.

The effective magnetic moment of a binuclear cluster is calculated by summing up the squares of separate moments $\mu^2(S')$ over all the spin layers with respect to a corresponding Boltzmann's factor. Then the effective magnetic moment of a binuclear cluster is expressed by Eq. (1.15):

$$\mu^2_{\text{eff}} = \frac{g^2}{n}\frac{\sum_{S'} S'(S'+1)(2S'+1)\exp[-E(S')/kT]}{\sum_{S'}(2S'+1)\exp[-E(S')/kT]}, \tag{1.15}$$

where n is number of magnetic ions in cluster.

Let us consider the binuclear cluster $[Cr_2(CH_3COO)_4(H_2O)_2]$, where Cr^{2+} ions have four unpaired electrons each and are located at the distance 2.64 Å. The experiment gives the effective magnetic moment increasing as the temperature increases and being lower than the spin only value. What information can be derived from these data? In this case $S_1 = S_2 = 2$, so that the quantum number S' will take the values 0, 1, 2, 3, 4. For every one of five levels the energy of exchange interactions $E(S')$ may be calculated from Eq. (1.14) and plotted as the so-called correlation diagram of the spin states (Fig. 1.3), showing the manner in which $E(S')$ changes upon changing the sign and value of the exchange parameter.

If the interaction of Cr–Cr atoms is antiferromagnetic, i.e., the spins of interacting atoms tend to antiparallel ordering ($J_{12} < 0$), at 0 K the compound must be diamagnetic ($S' = 0$). If the interaction is ferromagnetic ($J_{12} > 0$), the magnetic moment calculated per one Cr^{2+} atom at 0 K must be 6.33 μB. The temperature dependence of μ_{eff} may be calculated by summing up the values of $\mu^2_{\text{eff}}(S')$ by all the levels by Eq. (1.15) (Fig. 1.3). For the case under study we obtain the following formula (where $x = J_{12}/(kT)$):

$$\mu^2_{\text{eff}} = 2\frac{180e^{8x} + 84 + 30e^{-6x} + 6e^{-10x}}{9e^{8x} + 7 + 5e^{-6x} + 3e^{-10x} + e^{-12x}}$$

or

$$2\frac{180e^{20x} + 84e^{12x} + 30e^{6x} + 6e^{2x}}{9e^{20x} + 7e^{12x} + 5e^{6x} + 3e^{2x} + 1}.$$

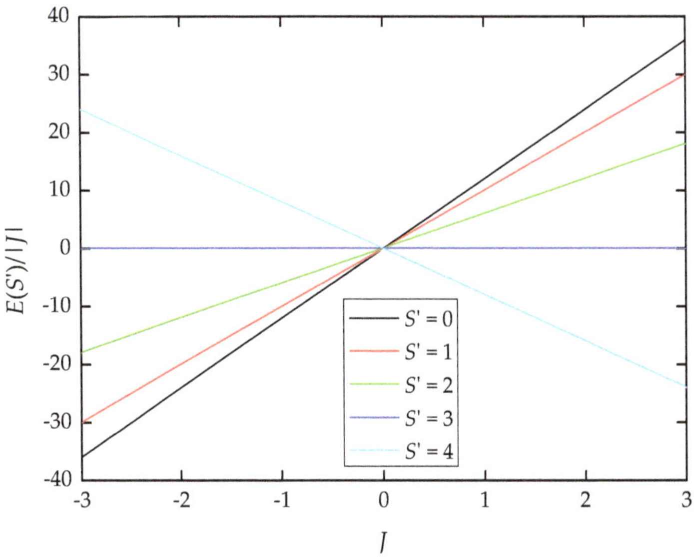

Figure 1.3 Spin states diagram for binuclear cluster $d^4 - d^4$ with various values of total spin S'.

Therefore, on the basis of experimental dependence of the effective magnetic moment of a binuclear cluster on temperature the value of exchange parameter can be determined. The magnetic properties of binuclear clusters as a function of the so-called adjusted temperature ($kT/|J|$) are shown in Fig. 1.4. For clusters with antiferromagnetic exchange, the effective magnetic moment increases as the temperature increases (since the levels with

$S' > 0$ become occupied), and for ferromagnetic clusters, it decreases.

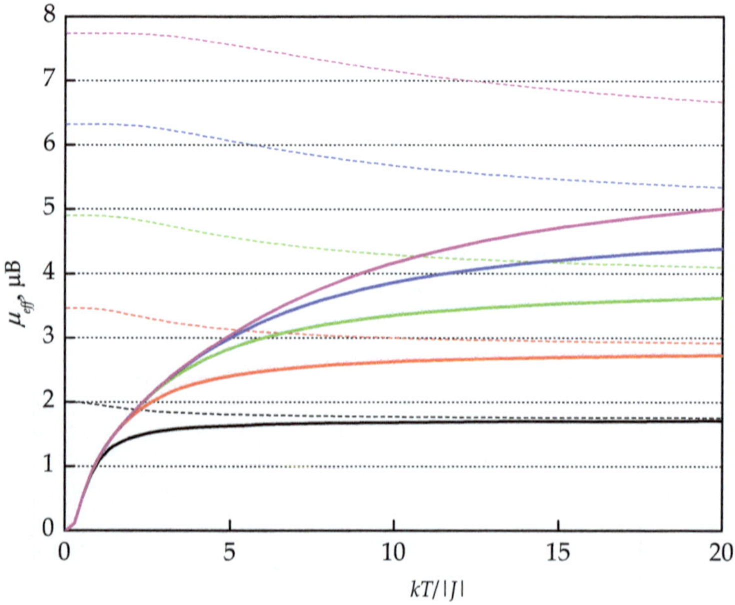

Figure 1.4 Temperature dependencies of effective magnetic moment for binuclear clusters with various J and equal spins: $J < 0$ (solid lines); $J > 0$ (dashed lines); $S_1 = S_2 = 1/2$ (black), 1 (red), 3/2 (green), 2 (blue), 5/2 (magenta).

Thus, from the experimental dependence of the effective magnetic moment the value and sign of the exchange parameter may be determined. For example, for the chromium clusters under study $J = -560\ \text{cm}^{-1}$.

For dimer clusters with various spins the equations for effective magnetic moment calculations may be the following according to Heisenberg–Dirac–Van Vleck model (Table 1.4).

The Heisenberg–Dirac–Van Vleck model may be extended to clusters with any number of atoms. However, in distinction to binuclear clusters we must sum over all the possible interactions between neighboring atoms. In this case for multinuclear clusters the exchange Hamiltonian assumes the form:

$$H = -2\sum_{i,j} J_{ij} S_i S_j. \tag{1.16}$$

Table 1.4 Expressions for effective magnetic moment for dimer clusters

Spins of magnetic centers	Expression for effective magnetic moment
$S_1 = S_2 = 1/2$	$\mu_{\text{eff}}^2 = 2\frac{6e^{2x}}{3e^{2x}+1};$
$S_1 = S_2 = 1$	$\mu_{\text{eff}}^2 = 2\frac{30e^{6x}+6e^{2x}}{5e^{6x}+3e^{2x}+1};$
$S_1 = S_2 = 3/2$	$\mu_{\text{eff}}^2 = 2\frac{84e^{12x}+30e^{6x}+6e^{2x}}{7e^{12x}+5e^{6x}+3e^{2x}+1};$
$S_1 = S_2 = 2$	$\mu_{\text{eff}}^2 = 2\frac{180e^{20x}+84e^{12x}+30e^{6x}+6e^{2x}}{9e^{20x}+7e^{12x}+5e^{6x}+3e^{2x}+1};$
$S_1 = S_2 = 5/2$	$\mu_{\text{eff}}^2 = 2\frac{330e^{30x}+180e^{15x}+84e^{12x}+30e^{6x}+6e^{2x}}{11e^{30x}+9e^{15x}+7e^{12x}+5e^{6x}+3e^{2x}+1};$
$S_1 = 1/2, S_2 = 1$	$\mu_{\text{eff}}^2 = 2\frac{30e^{3x}+3}{8e^{3x}+4};$
$S_1 = 1/2, S_2 = 3/2$	$\mu_{\text{eff}}^2 = 2\frac{30e^{4x}+6}{5e^{4x}+3};$
$S_1 = 1/2, S_2 = 2$	$\mu_{\text{eff}}^2 = 2\frac{105e^{5x}+30}{12e^{5x}+8};$
$S_1 = 1/2, S_2 = 5/2$	$\mu_{\text{eff}}^2 = 2\frac{84e^{6x}+30}{7e^{6x}+5};$
$S_1 = 1, S_2 = 3/2$	$\mu_{\text{eff}}^2 = 2\frac{105e^{8x}+30e^{2x}+3}{12e^{8x}+8e^{3x}+4};$
$S_1 = 1, S_2 = 2$	$\mu_{\text{eff}}^2 = 2\frac{84e^{10x}+30e^{4x}+6}{7e^{10x}+5e^{4x}+3};$
$S_1 = 1, S_2 = 5/2$	$\mu_{\text{eff}}^2 = 2\frac{252e^{12x}+105e^{5x}+30}{16e^{12x}+12e^{5x}+8};$
$S_1 = 3/2, S_2 = 2$	$\mu_{\text{eff}}^2 = 2\frac{252e^{15x}+105e^{8x}+30e^{3x}+3}{16e^{15x}+12e^{8x}+8e^{3x}+4};$
$S_1 = 3/2, S_2 = 5/2$	$\mu_{\text{eff}}^2 = 2\frac{180e^{18x}+84e^{10x}+30e^{4x}+6}{9e^{18x}+7e^{10x}+5e^{4x}+3};$
$S_1 = 2, S_2 = 5/2$	$\mu_{\text{eff}}^2 = 2\frac{495e^{24x}+252e^{15x}+105e^{8x}+30e^{3x}+3}{20e^{24x}+16e^{15x}+12e^{8x}+8e^{3x}+2}.$

It must be emphasized, however, that the Heisenberg–Dirac–Van Vleck model is applicable only to orbitally non-degenerated (A) or doubly degenerated (E) ground states of interacting atoms, since these states are not split by the spin–orbit coupling. The influence of spin–orbit coupling may be taken into account in this case by introducing the g-factor differing from 2 into Eq. (1.15).

The methods of calculating the exchange parameters for triplet ground states are described in ref. 4. The methods of calculating the exchange parameters for three and four nuclear clusters are given in details in refs. 2 and 4.

The examination of the data on the exchange parameters obtained on the basis of magnetic characteristics for various compounds of the same transition element, for example chromium, shows that they differ substantially. To understand the reasons for the exchange parameters variation in various compounds and associate them with electron structure and characteristics of the chemical bonds we must consider the possible channels of the exchange.

For the exchange effect to emerge it is necessary that the unpaired electrons of two atoms overlap in some manner. As has been noted above in the acetate binuclear complex of Cr^{2+} the distance between paramagnetic centers is rather small—only 2.64 Å. In this case d-orbitals of the neighboring atoms can overlap; such an exchange is called direct. In the coordination chemistry and in the solid-state chemistry the most commonly encountered are the cases, when the orbitals of paramagnetic atoms overlap with the orbitals of bridge ligands. Then such an exchange is called an indirect or superexchange interaction.

The exchange effect may be of two types: (1) if the orbitals of unpaired electrons overlap immediately or through the ligand, the exchange appeared to be antiferromagnetic (an analogy with the formation of the chemical bond, but with lower energy, $J < 0$); (2) if the interacting orbitals are orthogonal or overlap with orthogonal orbitals of a bridge ligand, a ferromagnetic exchange interaction emerges (an analogy with Hund's rule for free atoms, $J > 0$).

In Fig. 1.5 some examples are given of antiferromagnetic exchange. Figure 1.5a demonstrates the result of direct overlapping of the orbitals of interacting atoms (direct exchange). The case in Fig. 1.5b shows the superexchange at the angle 180° between two $d_{x^2-y^2}$ orbitals of a d-element via p_x-orbital of a ligand, i.e., $d_{x^2-y^2} \,||\, p_x \perp p_y \,||\, d_{xy}$. In this case the overlapping is strong and $J << 0$. Figure 1.5c shows what happens upon a weaker π-overlapping of d-orbitals with the orbitals of a ligand, i.e., $d_{xy}\,|p_y|\, d_{xy}$, $J < 0$. The same is valid for d_{xz} orbitals.

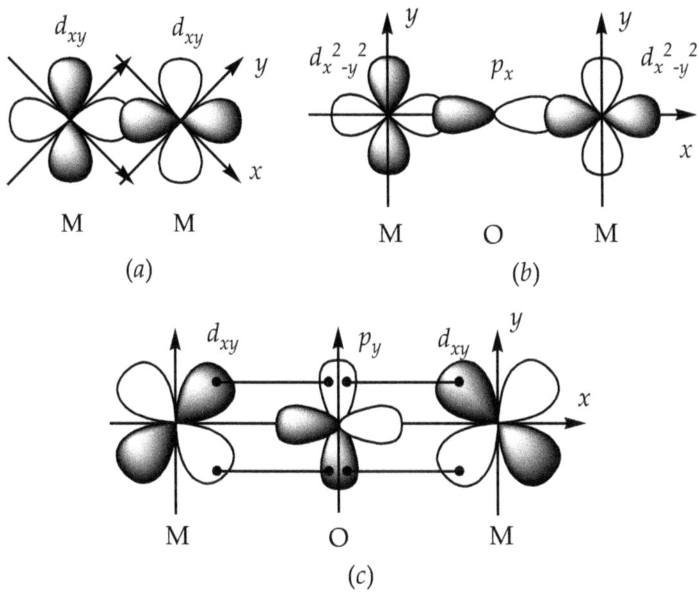

Figure 1.5 Schematic representation of direct overlapping of d-orbitals (a) and variants of superexchange via p-orbitals of oxygen atoms (b, c).

In the same cluster the ferromagnetic exchange occurs, for example, by the channels of the type $d_{xy} \,||\, p_y \perp p_z \,||\, d_{xz}$, $d_{x^2-y^2} \,||\, p_x \perp p_y \,||\, d_{xy}$. In this case $J > 0$. The mechanisms of the exchange are described in more details in refs. 2 and 7–9.

If there can be several exchange channels the following rule may be used: the exchange parameter J in the Hamiltonian of Heisenberg–Dirac–Van Vleck is the sum of the contributions over all the interatomic single electron exchange channels J_{ij} divided into the number of such channels equal to $4S_iS_j$. In this case the single electron contribution only slightly depends on the number of unpaired electrons of paramagnetic atoms.[7] Therefore, the model of exchange channels may be useful in magnetochemical practice when studying a series of isostructural compounds including various paramagnetic atoms, and also for approximate estimation of possible total exchange parameter in the clusters of various types.

So we can see how much information about the electron structure of complex compounds can be derived from magnetochemistry. Now let us see, how this method can be applied to solids, which are the basis for the most part of materials required by industry nowadays.

1.3 Magnetic Phenomena in Solids

In solid-state chemistry the method of magnetic susceptibility has rather long been in use, but its successes may be considered fairly modest. Let us try to make sense of this.

The exchange interactions in the coordination compounds take place far from always. In solids it is quite different. In the crystals paramagnetic atoms are embedded at the sites in the crystal lattice and inevitably have the nearest paramagnetic neighbors within even the same unit cell. This results in exchange interactions embracing a large number of atoms—long order or cooperative effect. The interactions may be direct (metals) or indirect (oxides, halides) as in the paramagnetic clusters.

The cooperative effect can result in a magnetic ordering—ferro-, ferri- or antiferromagnetic. At high temperatures such compound behave as paramagnetics, but as the temperature decreases, the spins of interacting paramagnetic atoms tend to ordering either in the same direction along the magnetic field or alternate in opposite directions. The type of magnetic ordering depends on the sign of the exchange parameter. The run of temperature dependences of magnetic susceptibility for various ordering is given in Fig. 1.6.

Ferromagnetic ordering of the spins is observed in some metals (Fe, Co, Ni) and in a series of complex compounds (CrO_2, γ-Fe_2O_3). At the temperatures higher than Curie point (T_C) the susceptibility of a ferromagnetic is described by Curie–Weiss law with positive Weiss constants. Lower than Curie point the susceptibility abruptly increases and appears to depend on the field strength. This effect is defined by a spontaneous splitting of the sample to small (~1 μm) regions of spontaneous magnetization—domains. The mutual orientation of domains is such that in the absence of magnetic field the total magnetization is zero. As the field is switched on the volume of the domains oriented close to the direction of magnetic field increases. In this case, the vector of spontaneous magnetization turns in the direction of magnetic field, and a sample gets magnetized. Upon switching off the magnetic field the spins are not disordered under the action of thermal motion only, a hysteresis loop is observed. To remove the ordering a field in the opposite direction must be applied.

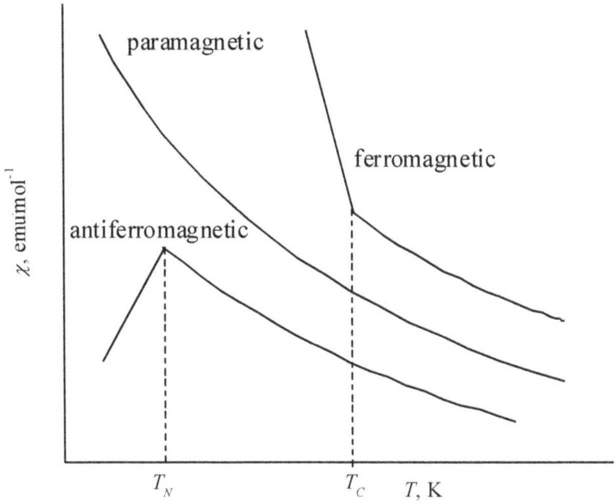

Figure 1.6 Temperature dependencies of magnetic susceptibility for magnetic ordering systems.

In a number of systems at a certain temperature called Neel temperature (T_N) the atom magnetic moments get ordered in such a manner that the spontaneous magnetization becomes zero. Such substances are called antiferromagnetics. Among them are many oxides of transition elements. In the simplest case, antiferromagnetic state may be sketched as magnetic moments alternating in their direction, i.e., an antiferromagnetic represents two equivalent ferromagnetic sublattices oriented antiparallel. At the temperatures higher than T_N the magnetic susceptibility obeys Curie–Weiss law with negative Weiss constants.

If the magnetic sublattices in a substance are nonequivalent or there are more than two sublattices, as, for example in the spinel Fe_3O_4, than their antiparallel ordering may result in ferrimagnetism. The magnetic moment of such compounds may be very high.

Magnetic ordering in crystalline solids imposes limitations on the method of magnetic susceptibility. However, even in the absence of magnetic ordering or when T_N or T_C are very low, as, for example in complex oxides, the interpretation of experimental data on magnetic susceptibility appears very complicated. The reason is that magnetic susceptibility χ_A appears to be the function depending not only on the spin state of a transition element, but on the character and energy of interatomic exchange interactions. An additional difficulty is that there are several directions of magnetic exchange in a crystal structure differing by their character and energy. For example, in the face centered cubic structure of NiO each nickel atom interacts directly with 12 nearest neighbors. In addition, it interacts with the same neighbors via oxygen atoms at an angle 90° and also with six neighbors at an angle 180°. We must not forget the translation symmetry of a crystal which also contributes into the total exchange. Thus, we have a great number of exchange parameters in solids, which are very difficult to distinguish. The situation is aggravated by the fact that whereas in complex compounds the parameters of the state of single atoms are comparatively easy obtained from electron spectra, in the solids this problem is seldom solved unambiguously due to the exchange broadening of the lines even in single crystals. The use of parameters determined for complex compounds is rarely rewarding since, for example, spin–orbit coupling appears to be partially frozen, and solids with infinite number of interacting atoms have a continuous spectrum of magnetic excitations,[2,4] thus the tabulated data for single electron constants are overstated. This resulted in the fact that the examination of magnetic susceptibility for paramagnetic crystals is half empirical and based on the temperatures of magnetic ordering in the framework of molecular field theory[10]:

$$T_O = \frac{2z|J|}{3k} S(S+1), \tag{1.17}$$

where T_0 is the temperature of magnetic ordering (Curie or Neel) in K, z the coordination number, S the spin of a paramagnetic atom, and J the parameter of ferro- or antiferromagnetic exchange in cm^{-1}.

Rushbrook and Wood method developed for high-temperature ($T > T_C$) region of the χ vs T for ferromagnetics[11] suggests the expansion of the susceptibility into the series by the degree of ($|J|/kT$). This method is successively used also for antiferromagnetics obtaining in some cases reasonable variations of $|J|$ for a series of similar compounds, for example $MCrO_2$ (M = Li, Na, K).[12]

However, it must be kept in mind that the exchange parameter determined by the temperature of magnetic ordering (if it is observed) and also by any other characteristics of magnetically concentrated systems is a certain average value, which does not reflect possible differences in the exchange along various directions, with the nearest neighbors and over the whole lattice. In other words it does not give a chemist any insight into the nature of the short order interactions between paramagnetic atoms.

The attempts to expanse the magnetic susceptibility of magnetically concentrated systems at $T > T_C$, i.e., in the region, where Curie–Weiss law is conventionally obeyed, resulted in examination of diluted systems, i.e., solid solutions. Their susceptibility was considered by Blocker and West[13] within the framework of the same molecular field theory. In this case, the exchange interactions were included into Weiss constant, which resulted in formula

$$\frac{aC}{||\chi||} = T + a\theta, \tag{1.18}$$

where a is the mole fraction of a paramagnetic, $||\chi||$—tensor of magnetic susceptibility; C the Curie constant, $C = \frac{Ng^2\beta^2}{3k}S(S+1)$; and θ the Weiss constant, $\theta = \frac{1}{3k}2S(S+1)\sum_i^m z_i J_i$.

The main drawback of this method is the assumption that the distribution of paramagnetic atoms in the lattice of the solid solution is statistically disordered. This pertains equally to a majority of authors, who in one way or other deal with solid solutions (see, for example, ref. 14). As will be shown below, the statistically disordered distribution is an exception rather than a rule. The second very important obstacle consists in the following. If we compare fundamental Curie equation for magnetic susceptibility of a paramagnetic (Eq. 1.2) with empirical Curie–Weiss law, it becomes evident that in Eq. (1.3) the exchange interactions are embedded into the temperature dependence of the effective magnetic moment. However, what is very important, this very dependence includes such characteristics as spin–orbit coupling, bond covalence, etc. If as the first approximation the latter may be neglected, for the triplet ground states T split by spin–orbit coupling there is an abrupt dependence of the effective magnetic moment on temperature.[1–4] In the experiment this gives a nonzero θ, which is not associated with exchange interactions. Therefore, strictly speaking, the use of the above given approach may provide an adequate information only for a very limited number of systems.

When the problems of synthesis of compounds with predetermined complex of physical and chemical properties arise, and in the solid-state chemistry these are the problems of selecting the composition of multicomponent systems, the structure, and the thermal treatment regimes. In this case, one of the most important aims of a chemist is to have a grasp of the chemical bond at the short range level, i.e., about the state of A atoms, their interactions with neighboring atoms, since this allows the predictions to be made about the influence of replacing the atoms in the nearest surrounding of A atom on its characteristics, and, consequently, on the properties of a target material in the end.

1.4 Magnetic Dilution Method

It appears tempting to find a method, which could apply a very well developed theoretical grounds of magnetic susceptibility to reveal the electron structure in solids. With this purpose it is necessary to separate in some manner the solid into single atoms and groups of atoms. This initiated the development of the method of magnetic dilution. The aim of this method is to obtain an exhaustive information about the state of paramagnetic atoms in oxide systems, about the character and energy of exchange interactions, and about the influence of the composition and structure on these parameters. The essence of the method is the study of magnetic susceptibility of diluted solid solutions of isomorphous substitution of paramagnetic substance in a diamagnetic matrix. It must be emphasized that the majority of materials used nowadays, such as colossal magnetoresistors, electrodes for solid oxide fuel cells (SOFC) and solar cells, as catalysts and fluorescent systems are multicomponent systems and do represent solid solutions.

From the point of view of thermodynamics according to the data of refs. 15 and 16, solid solutions of oxide systems are regular solutions, i.e., the systems with a small, but as a rule nonzero mixing enthalpy, which is related to the interchange energy

$$\omega_{12} = H_{12} - ½(H_{11} + H_{22}), \tag{1.19}$$

where H_{ij} is the energy of interaction between like and unlike atoms in the solid solution.

This is important in the process of considering magnetic properties, since the distribution of the atoms is strictly statistical only in the ideal solutions. Any deviation from the ideal inevitably results in a certain segregation. Here is the reason for inadequacy of using the statistical distribution for calculating the susceptibility of magnetically diluted systems. However, low values of ω must result in the fact that at a considerable dilution the entropy factor, which is great because the synthesis of solid solutions is usually carried out at high temperatures, plays increasingly greater role, and in the limit, at an infinite dilution this must result in single atoms of the dissolved component only being present in the system. In such a case the extrapolation of magnetic characteristics to zero concentration of the paramagnetic atoms may give information about the state of single *d*-element atom in the same coordination surrounding as in the pure compound. In general, for the structures, where the parameters of the nearest surrounding do not change, only a minor change in the volume of coordination polyhedron is observed, we can be assured that magnetic characteristics at $x \to 0$ reflect the state of paramagnetic atom in the structure of pure compound. Hereafter we shall consider oxide systems, but there is no principal limitation on the compound types for the use of this method.

The temperature dependences of magnetic characteristics give the possibility of estimating the distribution of paramagnetic atoms over the particular lattice and of the exchange interactions between them. However, the method of magnetic dilution imposes rigid conditions on the synthesis of solid solutions and their characterization. Oxide materials described in abundance in the current literature are for the most part solid solutions, but their concentrations under study are usually scarce. Magnetic dilution method requires obtaining a series of 8–10 compositions in the concentration range 1–10 mol% of paramagnetic element. The solutions obtained by any available synthetic procedure (ceramic, sol-gel, etc.) must be thoroughly studied by X-ray method. However, if the obtained solutions are found single phase by X-ray method, this is not enough, since the distribution of paramagnetic atoms must be as close to the equilibrium, as possible. For this purpose, the samples must be sintered until their magnetic susceptibility, which is the function of atom distribution, becomes constant. If possible, it is worthwhile to increase the sintering temperature for a time by about 50–100°C and then decrease it back and sinter for the same time—approaching the equilibrium from two sides. The constancy of magnetic susceptibility proves that the obtained samples are equilibrium and suitable for thermodynamic estimations. Then it is necessary to make a chemical analysis to determine the content of paramagnetic atoms, which can change in the process of high-temperature treatment.

The magnetic susceptibility of the samples is measured over a widest possible temperature range. The diamagnetic matrix must be measured over the same temperature range since it is necessary to introduce the diamagnetic corrections. For diluted solutions, these diamagnetic corrections may be large enough. Thus the paramagnetic component of magnetic susceptibility for the $LaM_xAl_{1-x}O_3$ solid solution taken as the example will be calculated by Eq. (1.20):

$$\chi_M = \frac{M^x \chi_g^x - (1-x)M^O \chi_g^O}{x} - \sum \chi_i^{dia}, \tag{1.20}$$

where χ_g^x, and χ_g^0 are specific magnetic susceptibilities of the solid solution and diamagnetic matrix ($LaAlO_3$ in this example), respectively, in emu·g^{-1}; M^x and M^o their molar masses in g·mol^{-1}; $\Sigma\chi_i^{dia}$ the sum of tabulated data on diamagnetic corrections for the dissolved substance ($LaMO_3$) in emu·mol^{-1}; and x the concentration of paramagnetic atoms in the solid solution.

Now let us see what information can be derived from magnetic susceptibility data for diluted solid solutions.

Having measured magnetic susceptibility of several (about 8–10) samples of solid solutions over a wide temperature range and at 16–20 fixed temperatures we can plot the isotherms of paramagnetic susceptibilities and temperature dependences of inverse χ_M and of effective magnetic moments. The number of experimental points allows a reliable extrapolation of magnetic characteristics to the infinite dilution ($x \to 0$), thus separating the problem of chemical structure determination into two parts—the electron structure of single paramagnetic atoms, and the study of exchange interactions as they develop with the increase in concentration. Within such an approach all the fundamentals developed in the magnetochemistry of coordination compounds can be successfully used.

For *d*-elements with *A*- and *E*-ground states, their μ_{eff} being independent on temperature, for example Cr^{3+} ($^4A_{2g}$), Ni^{2+} ($^3A_{2g}$), Cu^{2+} (^2E_g) in the octahedral field by the value of μ_{eff} it is possible to determine the ratio $\lambda/10Dq$ from Eq. (1.9). If from some independent experiments $10Dq$ are known, the spin–orbit coupling constant can be determined. For the $Cu_xMg_{1-x}Al_2O_4$ solid solutions[17] λ was found to be −225 cm^{-1}. For the triplet ground states, where $\mu_{eff} = f(kT/\lambda)^1$ λ for V^{3+} in $LaV_xAl_{1-x}O_3$ was found to be 30 cm^{-1}. This proves the statement of ref. 2 about a partial freezing the spin–orbit coupling in the solids—all the absolute values of spin–orbit coupling constants appear to be less than in complex compounds (see Table 1.2).

However it must be taken into account that μ_{eff} for orbitally non-degenerated ground term does not depend on temperature only in the case of regular octahedral surrounding. A distortion of the octahedron results in the so-called zero field splitting of the ground term and in some dependence of μ_{eff} on temperature.[18] The value of this splitting as well as the energy interval between the ground and excited terms also may be estimated from the dependence of μ_{eff} vs T, which was carried out for the $Ni_xMg_{1-x}Sb_2O_6$ solid solutions with trirutile structure.[19] In this case, the obtained data—$D = 7\ cm^{-1}$, $\Delta = 319\ cm^{-1}$, where D is the zero field splitting, Δ the energy difference between $^3A_{2g}$ and $^3T_{1g}$ terms—appeared to be in good agreement with the data of optic studies carried out simultaneously. It is interesting to note that $D = 7\ cm^{-1}$ fits a rather essential distortion of the octahedron—z axis is at the angle 78° to the basis plain.

From the above-mentioned example, it follows that upon structural phase transitions in the oxide lattice the changes in the symmetry of the nearest oxygen surrounding of a transition metal must result in a change in the $\chi_M - T$ ($\mu_{eff} - T$) dependences. In such a case transition element atoms introduced into a diamagnetic matrix may serve as a probe for revealing structural phase transitions. This can be proved by the example of the $Sr_2M_xMg_{1-x}WO_6$ system with double perovskite structure. Upon substituting magnesium, atoms for Co^{2+} and Ni^{2+} in the Sr_2MgWO_6 matrix there appeared breaks in the dependences of inverse paramagnetic susceptibility on temperature resulting in the change in the plots of μ_{eff}–T this testified for a phase transition in the temperature range 220–270 K, resulting in a decrease in the symmetry.[20]

Magnetic properties of complex oxides containing transition elements with the electron configuration d^5, d^6, d^7 - Co^{3+}, Ni^{3+}, Ru^{3+}, Rh^{3+} often appear to be complicated for interpretation. This is associated with the possibility of various spin states—high-spin and low-spin; for Co^{3+} there can be even intermediate spin $S = 1$. The coordination chemistry used the theory of spin states in such cases.[6] This theory was shown to be applicable to the complex oxide with perovskite structure for μ_{eff} extrapolated to zero concentration in $LaAlO_3$. The spin transition energies obtained in this manner are in good agreement with published spectra data for $LaCoO_3$ and in a logic manner describe the tendencies of spin transition energy variation in the Periodic Table[21] (Fig. 1.7).

In a number of crystal structures, such as spinels, garnets, etc., there are nonequivalent sites occupied by the metal atoms—in spinel, for example, it is octahedral and tetrahedral sites. The problem is that when dealing with multicomponent oxide system the interpretation of X-ray data on the distribution of a *d*-element over two different sites may be ambiguous, whereas the properties of the materials based on such systems may depend to a great extent on this distribution. Here again the fundamentals of the theory developed for complex compounds may be used successfully enough (see Eq. 1.11 and Table 1.3).

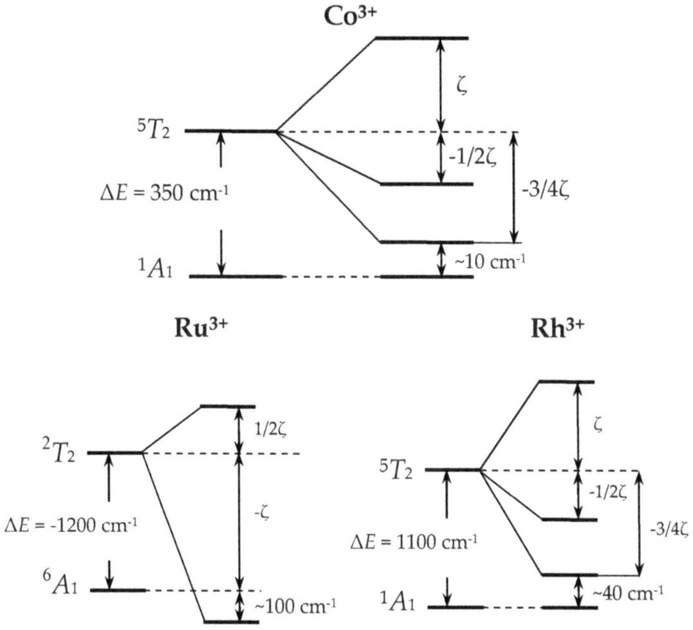

Figure 1.7 Schematic representation of energy of low- and high-spin transitions for Co^{3+}, Ru^{3+} and Rh^{3+} (ΔE is energy of crystal field).

In Fig. 1.8 the experimental dependence of the effective magnetic moment on temperature is given for the infinitely diluted $Cu_xMg_{1-x}Al_2O_4$ solid solution and theoretical dependences for Cu^{2+} in tetrahedral and octahedral surrounding. The fraction of copper atoms in the tetrahedral sites appeared to be 0.5.[17]

Therefore, after obtaining exhaustive information about the electron states of single paramagnetic atoms of *d*-elements in particular complex oxide structure the next problem is to describe the interatomic interactions in the lattice.

The essence of magnetic dilution method is that in the diluted solid solutions there exist single paramagnetic atoms and their small aggregates or clusters bonded by exchange interactions and having no paramagnetic atoms in their close vicinity. Then the susceptibility of such a solid solution may be expressed by Eq. (1.21):

$$\chi_M = \sum_i a_i \chi_i, \tag{1.21}$$

where a_i is the fraction of clusters with the number of atoms i (e.g., i = 1 means single paramagnetic atom, i = 2 means dimer, etc.) and χ_i magnetic susceptibilities of clusters in emu · mol^{-1}.

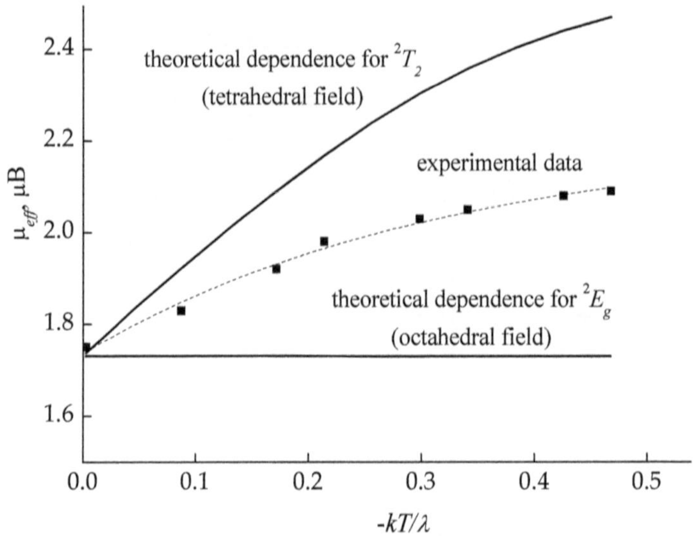

Figure 1.8 Experimental values of effective magnetic moment for $Cu_x Mg_{1-x}Al_2O_4$ at the infinite dilution (x = 0) and theoretical dependencies for Cu^{2+} at tetrahedral and octahedral fields.

However, in very diluted solutions at $x \leq 0.05$ the clusters with $i > 2$ are usually unlikely, at least upon isovalent substitution, then Eq. (1.21) may be rewritten as

$$\chi_M = (1 - a_2)\chi_1 + a_2\chi_2.$$

In this case the susceptibility will be determined by two parameters: J_{12}—the same for all the dimers—and a_2—the dimer fraction monotonously increasing with the increase in concentration of paramagnetic atoms, but the same for a solid solution with particular x over the whole temperature range under study. A large series of concentration and temperature dependences of magnetic characteristics makes the calculation of these parameters reliable enough. Such calculations were carried out for a series of $LaM_xAl_{1-x}O_3$ solid solutions with cubic perovskite structure (M = Ti, V, Cr, Mn, Fe, Co, Ni). Important regularities were found for these solid solutions. First of all the distribution of paramagnetic atoms is far from statistical, the fraction of dimers is about twice of statistically probable (Fig. 1.9).

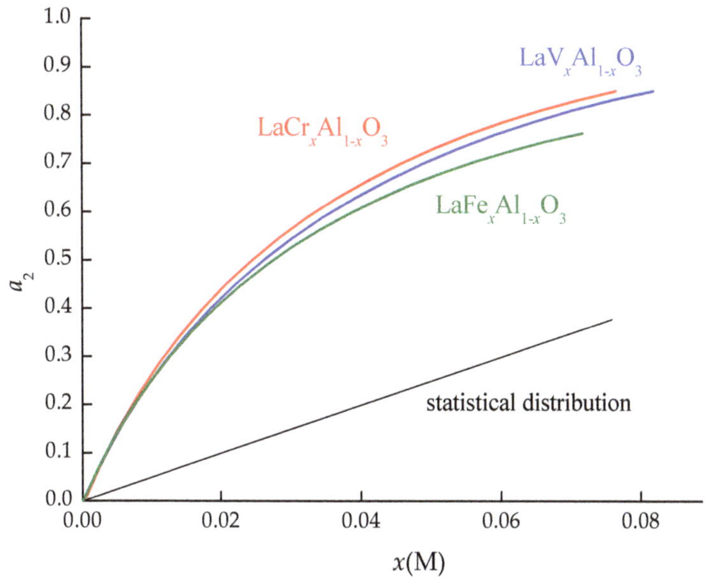

Figure 1.9 Dependence of the fraction of dimer clusters on paramagnetic atom concentrations for solid solutions based on lanthanum aluminate.

Second, whereas the exchange parameters for the elements under study vary in agreement with the exchange channel model,[2,7] the fraction of dimers in the solid solutions does not depend on the nature of transition element. This can be understood if we compare the energies of exchange interactions, which do not exceed 1 kJ·mol^{-1} in the absence of magnetic ordering, and the interchange energy of regular solid solution formation (Eq. 1.19), which, according to calculations based on the obtained a_2,[15] amounts to 10–20 $kJ \cdot mol^{-1}$ for perovskite-like complex oxides.

Another very important regularity obtained on studying the magnetic susceptibility of solid solutions is that it allows the influence of diamagnetic atoms on the clustering of paramagnetic atoms in the solid solutions to be revealed. In the study of complex oxides with perovskite-like K_2NiF_4 structure ($AA'MO_4$, where A and A′ are large diamagnetic cations, e.g., La, Y, Ca or Sr), $YCaCr_xAl_{1-x}O_4$, $LaSrCr_xAl_{1-x}O_4$, and $Sr_2Mn_xTi_{1-x}O_4$, containing isoelectronic Cr^{3+} and Mn^{4+} the exchange parameters varied in the sequence: $YCaCr_xAl_{1-x}O_4$ (J = −13 cm^{-1}) < $LaSrCr_xAl_{1-x}O_4$ (J = −18 cm^{-1}) < $Sr_2Mn_xTi_{1-x}O_4$ (J = −30 cm^{-1}). At the same time clustering in the solid solutions increased in the same sequence (Fig. 1.10).

These regularities may be explained on the basis of bond ionicity competition A–O–M in perovskite-like oxides. The ionicity of the Sr–O bond is higher than the average ionicity of (Sr, La)–O bond, and the latter is higher than the ionicity of (Y, Ca)–O bond, which is determined by polarizing effect of smaller La, Y, and Ca atoms on the orbitals of oxygen. Then the M–O bonds become more covalent on passing from pair (Y, Ca) to (Sr, Sr) in $Sr_2Mn_xTi_{1-x}O_4$, which increases the overlapping of *d*-orbitals of paramagnetic atom with *p*-orbitals of oxygen, thus increasing the antiferromagnetic exchange and also the interchange energy. This can be explained as an increase in the ionicity of A–O bond, the covalency of M–O bond, which results in an increase in $|J|$ and an increase in paramagnetic atom clustering in the series $YCaAlO_4 \rightarrow LaSrAlO_4 \rightarrow Sr_2TiO_4$ (Fig. 1.10).

From the examples mentioned above it is evident that the possibilities of the method of magnetic dilution are numerous and not yet exhaustively studied. There are two essential advices to be given. First, to make unambiguous conclusions and create reliable models of electron structure of solids it is necessary to study wide series of solid solutions with same structure and various structures, differing by the nature of transition elements and diamagnetic elements and by their ratio. Second, whenever possible other physical methods must be used to support the magnetic measurements.

Of course, first of all the structural studies must be carried out, the more so that X-ray method is now developed for powder samples (Rietveld refinement, the measurements over a large range of temperature, etc.). But also, when possible, ESR and Mössbauer spectra, Raman spectra, high temperature mass spectrometry, etc., can give additional information about the crystal and electron structure of the systems under study.

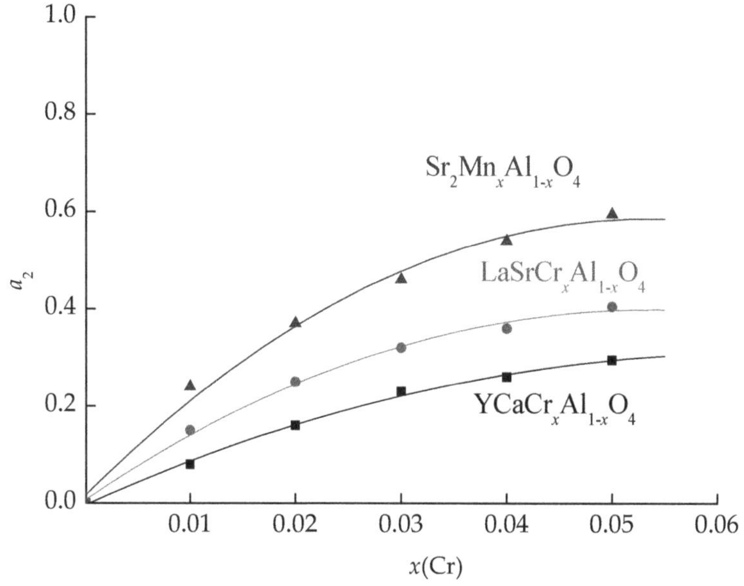

Figure 1.10 Dependencies of the fraction of dimer clusters on concentration of chromium or manganese for $YCaCr_xAl_{1-x}O_4$, $LaSrCr_xAl_{1-x}O_4$ and $Sr_2Mn_xTi_{1-x}O_4$ solid solutions.

For example, when studying the effect of substituting strontium for calcium in the $LaSrNi_xAl_{1-x}O_4$ solid solutions, the fraction of low-spin nickel was found to vary irregularly with calcium concentration by the data of magnetic susceptibility simulation. This fact found its explanation in the distribution of diamagnetic atoms in the lattice. However, there existed some ambiguity in such an irregularity. However, the ESR spectra gave the same dependence of low-spin nickel fraction in the solid solutions[22] thus supporting the suggested model of chemical structure. A large variety of new materials—electrodes for SOFC, magnetoresistors, catalysts, and so on—are based on heterovalent substitution in the well-known lattices. The introduction of dia- and paramagnetic elements often results in the stabilization of the target structure, in quite new electrical and magnetic performance. Therefore, the selection of the best suitable substitution (qualitative and quantitative) becomes most urgent. This selection must be made on the basis of the exhaustive knowledge of the changes in the electron structure. We shall present several examples of using the magnetic dilution method for revealing the electron structure of complex oxides promising for new materials and used as such nowadays.

References

1. Earnshaw A (ed) (1968) *Introduction to Magnetochemistry*, Acad Press, London, New York.
2. Kalinnikov VT, Rakitin YuV (eds) (1980) *Introduction to Magnetochemistry. Method of Static Magnetic Susceptibility in Chemistry*, Nauka, Moscow [In Russian].
3. Carlin RL (ed) (1986) *Magnetochemistry*, Springer Verlag, Berlin, Heidelberg, New York, Tokyo.
4. Rakitin YuV, Kalinnikov VT (eds) (1994) *Modern Magnetochemistry*, Nauka, St. Petersburg. [In Russian].
5. Van Vleck JH (1978) Quantum mechanics: The key to understanding magnetism, *Rev Mod Phys*, **50**, 181–190.
6. Martin RL, White AH (1968) The nature of the transition between high-spin and low-spin octahedral complexes of the transition metals, *Trans Met Chem*, **4**, 113–198.
7. Eremin MV, Rakitin YuV (1977) Channel model in isotropic exchange theory, *Phys. Stat Sol (b)*, **80**(4), 579–587, Ibid. (1977) **82**(2), 221–228; Ibid. (1978) **85**(6), 783–788.
8. Anderson PW (1959) New approach to the theory of superexchange interactions, *Phys Rev*, **115**(1), 2–13.
9. Anderson PW (1963) Exchange in insulators: Superexchange, direct exchange, and double exchange, in *Magnetism*, Acad. Press, New York, London, pp 25–83.
10. Krupicka S (1973) *Physik der ferrite und der verwangten magnetischen oxide*, Prag: Acad. verlag der Tschechlovakishen Akademie der wissenschaften.
11. Rushbrook GS, Wood PJ (1958) High temperature susceptibility of Heisenberg ferromagnetics, *Mol Phys*, **1**(3), 257–283.

12. Delmas C, Le Flem G, Fouassien C, Hagenmuller P (1978) Calcul des integrals d'echange, *J Phys Chem Sol*, **39**, 55–57.
13. Blocker TG, West FG (1969) On magnetic susceptibility of randomly diluted systems, *Phys Lett*, **28**, 487–488.
14. Van Leeuwen GCM (1973) Investigation of the angular dependence of Co–O–Co superexchange by means of magnetic susceptibility measurements in dilute systems, *Rec Trav Chim*, **92**, 1249–1266.
15. Smirnova NA (ed) (1973) *Metods of Statistical Thermodynamics in Physical Chemistry*, Vysshaya Shkola, Moscow. [In Russian].
16. Kozheurov VA (1975) *Statistical Thermodynamics*, Metallurgiya, Moscow [In Russian].
17. Brach BYa, Chezhina NV (1986) Magnetic estimation of copper(II) distribution in the $Cu_xMg_{1-x}Al_2O_4$ solid solutions with spinel structure, *Zh Obshc Khim*, **31**(8), 1957–1960.
18. Figgis BN (1960) Magnetic properties of transition metal ions in asymmetric ligand fields, *Trans Faraday Soc*, **56**, 1553–1557.
19. Brach BYa, Zvereva IA, Chezhina NV, Shchipunov EA (1987) Magnetochemical and spectral study of diluted solid solutions $Ni_x Mg_{1-x}Sb_2O_6$, *Zh Neorg Khim*, **32**(1), 29–33.
20. Brach BYa, Chezhina NV (1983) Magnetic properties of Fe, Co, Ni, Mn in the solid solutions $Sr_2Me_xMg_{1-x}WO_6$, *Izv AN SSSR Neorg Mat*, **19**(2), 292–294.
21. Brach BYa, Zvereva IA, Piir IV, Chezhina NV (1984) Study of the transition low-spin–high-spin in the Co(III), Ru(III), Rh(III) atoms in $LaAlO_3$ lattice, *Zh Neorg Khim*, **29**(6), 1387–1390.
22. Andronenko SI, Andronenko RR, Zagrebel'nyi OA, Chezhina NV (2002) States of atoms and interatomic interactions in complex perovskite-like oxides: XVI. Spin state of nickel(III) atoms in $Y_yLa_{1-y}CaNi_xAl_{1-x}O_4$ solid solutions by EPR and magnetic susceptibility, *Rus J Gen Chem*, **72**(12), 1853–1856.

Chapter 2

Phase Composition and Magnetic Characteristics of Solid Solutions and Complex Oxides Based on Scandium Molybdate $Sc_{2-2x}Gd_{2x}Mo_3O_{12}$ ($0 \leq x \leq 1$)

Dmitry A. Korolev and Mariia D. Sapova

Department of General and Inorganic Chemistry, St. Petersburg State University, Universitetskaya nab. 7/9, St. Petersburg, 199034, Russia

chemdim@mail.ru, d.korolev@spbu.ru

2.1 Introduction

Scandium molybdate is the first obtained stable material with negative coefficient of thermal expansion (CTE). The structure of scandium molybdate may be represented as a three-dimensional carcass of joining ScO_6 octahedra and MoO_4 tetrahedra (Fig. 2.1). An anisotropic thermal expansion occurs as the result of a decrease in the sizes along two axes at the expense of rotation of polyhedral parts and an increase along the third axis.[1]

Electronic Structure of Materials: Challenges and Developments
Edited by Natalia V. Chezhina and Dmitry A. Korolev

ISBN 978-981-4800-55-6 (Hardcover), 978-0-429-24287-8 (eBook)
www.panstanford.com

For scandium molybdate thermodynamically stable are monoclinic phase in the region of low temperatures and orthorhombic phase in the region of high temperatures.

In general the formation enthalpy of monoclinic phase is lower than of the orthorhombic phase; however, the orthorhombic phase is preferable from the point of view of entropy. This supports the fact that the monoclinic phase is stable at low temperatures and the orthorhombic phase—at high temperatures. In this case the phase transition is reversible. Upon the phase transition from orthorhombic to monoclinic phase an abrupt decrease in the volume of the unit cell is observed.

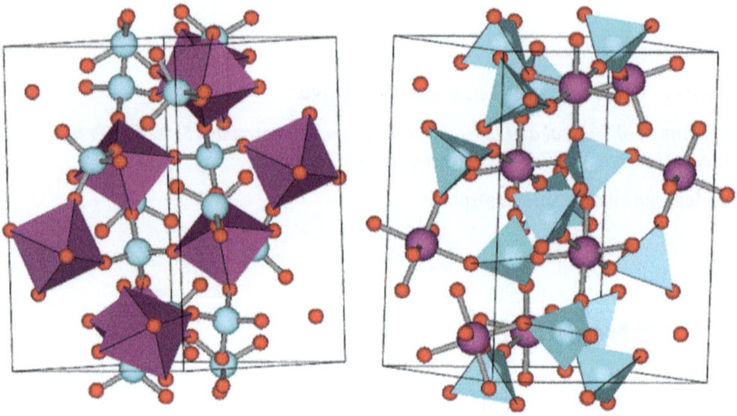

Figure 2.1 Joining the polyhedral in the structure of orthorhombic scandium molybdate: (Left) The octahedra occupied with scandium are given. (Right) Tetrahedral occupied with molybdenum.

The studies of gadolinium molybdate are not so comprehensive and are largely associated with optic properties of the crystals, which are ferrielectrics, i.e., showing spontaneous polarization up to a certain temperature (Curie temperature). According to ref. 2, up to T = 159°C gadolinium molybdate crystallizes in the orthorhombic syngony, and as the temperature increases a transfer to the tetragonal syngony is observed. The unit cell of the orthorhombic phase consists of 68 atoms; the tetragonal unit cell includes 34 atoms and is turned by 45° relative to the orthorhombic unit cell.

At the same time other researchers[3] advocate that orthorhombic ferroelectric phase of gadolinium molybdate is metastable. The formation of tetragonal phase occurs at 1165, and at 857°C the phase transition to monoclinic α-phase is observed. Since this transfer is rather slow, the high temperature may be preserved by rapid cooling. If the thermodynamically metastable tetragonal β-phase of $Gd_2Mo_3O_{12}$ is cooled further, it undergoes the second phase transition at 159°C. The result of this transformation is orthorhombic β_0-phase with a lower symmetry *Pba*2. This phase is also metastable thermodynamically, but it is very stable from the point of view of kinetics.

The review of published literature shows that the systems based on scandium molybdate have not been adequately explored. Hence, in this work an attempt was made to obtain structural, spectroscopic, and magnetic data of the systems based on scandium molybdate doped with gadolinium and to trace the evolution of phase formation in the $Sc_2Mo_3O_{12}$–$Gd_2Mo_3O_{12}$ system.

2.2 Synthesis, Structural and X-Ray Data

Complex oxides and solid solutions with the composition $Sc_{2-2x}Gd_{2x}Mo_3O_{12}$ (x = 0; 0.005; 0.025; 0.05; 0.075; 0.10; 0.25; 0.5; 1) were obtained by ceramic procedure by sintering the oxide mixture for 10 h at the temperature 900°C. The solutions with low concentration of gadolinium after the first sintering contained an admixture of molybdenum oxide, MoO_3, which was removed after the repeated grinding and sintering under the same conditions.

Given the concentrations x(Gd) > 0.10, the second phase is formed with the structure of gadolinium molybdate (GMO), which is seen in Fig. 2.2.

The samples of solid solutions were shown to be single phase and have the structure of scandium molybdate (SMO) in the concentration region x(Gd) ≤ 0.075. The examples of X-ray patterns obtained for single phase and two phase samples are given in Figs. 2.3, 2.4, 2.5.

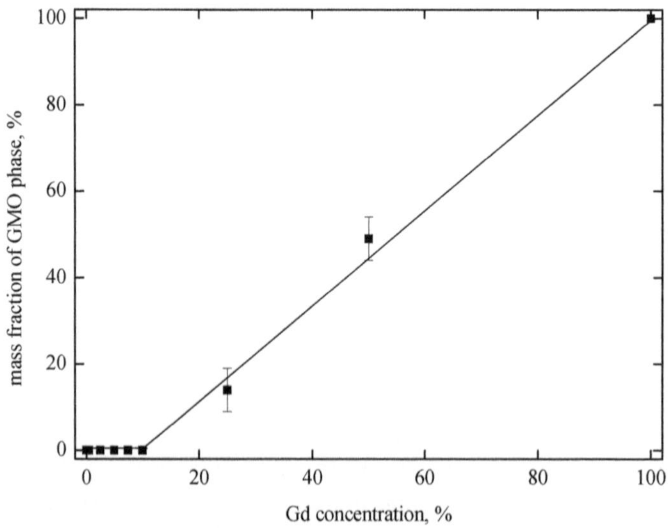

Figure 2.2 Plot of the mass fraction of the phase with GMO structure vs. gadolinium concentration in the system.

As the result of X-ray patterns calculation by the Rietveld method, we obtained the data about the phase composition of the samples, the unit cell parameters, and also the occupation of the sites by trivalent ions in various phases. The R_{wp} index being lower than 15% upon the analysis by the Rietveld method of the structures with low symmetry points to a good approximation of theoretical calculations to the obtained experimental data.

It can be seen from Table 2.1 that for the phase of scandium molybdate the unit cell parameters increase as the concentration of gadolinium increases to x(Gd) = 0.10. This can be accounted for by the introduction of increasing quantity of gadolinium atoms in to the structure, which have greater radius (0.938 Å) in comparison to scandium atoms (0.745 Å),[4] thus resulting in a broadening of the unit cell. Given the concentration x(Gd) = 0.10 a minor decrease in the unit cell parameters is observed, which may be associated with the formation of new phase of gadolinium molybdate, in spite of the fact that this phase was not registered in the X-ray patterns. However, the emergence of this phase can be fixed by the method of static magnetic susceptibility, as will be described below. This can be accounted for by an inadequate sensitivity of X-ray diffraction method.

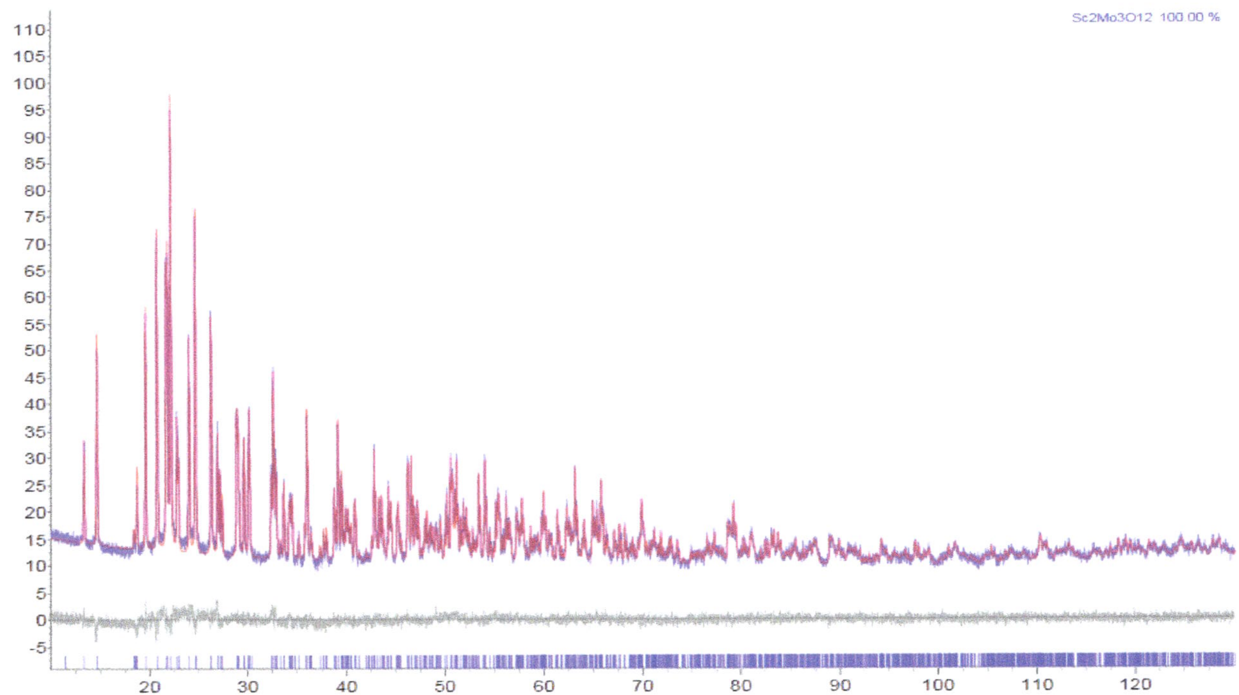

Figure 2.3 Rietveld refinement of the X-ray pattern for $Sc_{1.95}Gd_{0.05}Mo_3O_{12}$ (R_{exp} = 4.1%; R_{wp} = 6.9%; GOF = 1.7).

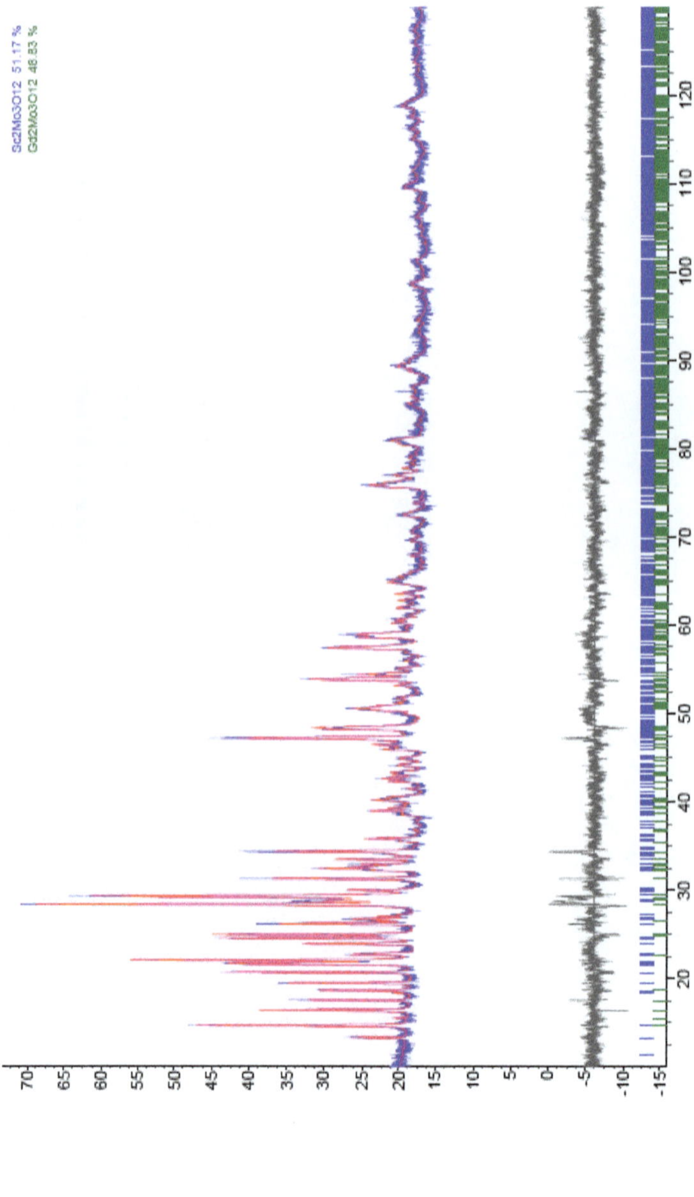

Figure 2.4 Rietveld refinement of the X-ray pattern for $Sc_{1.5}Gd_{0.5}Mo_3O_{12}$ (R_{exp} = 3.1%; R_{wp} = 7.9%; GOF = 2.5).

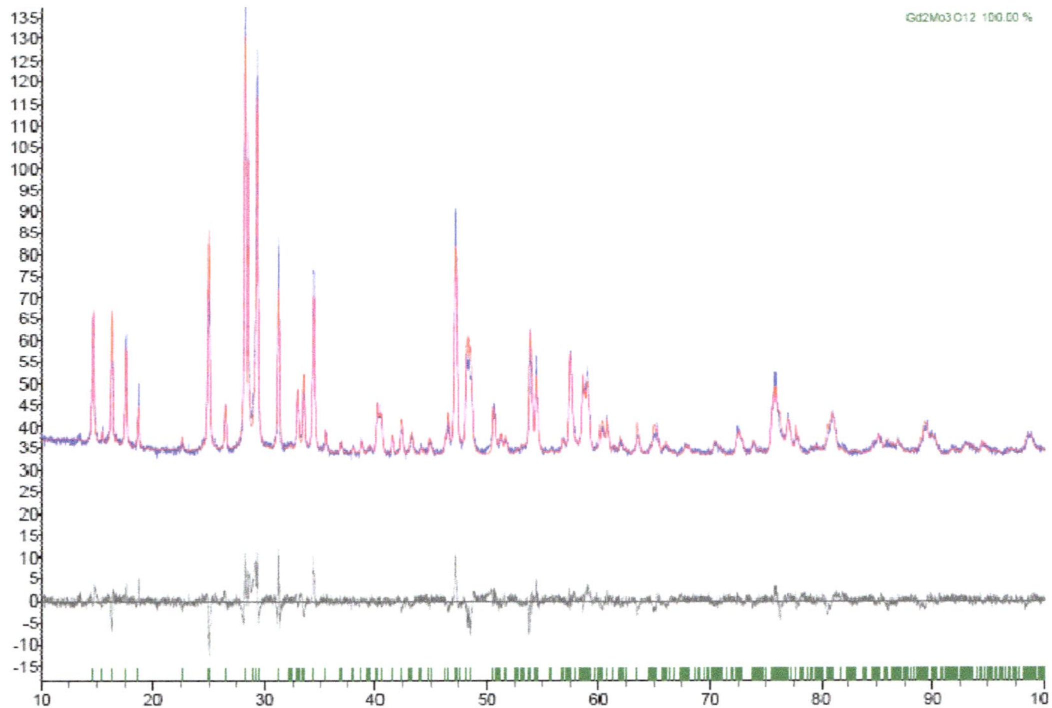

Figure 2.5 Rietveld refinement of the X-ray pattern for $Gd_2Mo_3O_{12}$ (R_{exp} = 6.5%; R_{wp} = 8.6%; GOF = 1.3).

Table 2.1 Unit cell parameters for solid solutions and complex oxides by Rietveld refinement

	Unit cell parameters with SMO structure (Å)			Unit cell parameters with GMO structure (Å)				R_{wp}
x(Gd)	a	b	c	a	b	c	β,°	(%)
0	13.2482	9.5478	9.6409	—	—	—	—	9.4
0.005	13.2518	9.5491	9.6421	—	—	—	—	11.7
0.025	13.2635	9.5527	9.6464	—	—	—	—	8.6
0.05	13.2762	9.5558	9.6472	—	—	—	—	6.2
0.075	13.2928	9.5640	9.6604	—	—	—	—	9.2
0.10	13.2800	9.5584	9.6497	—	—	—	—	6.7
0.25	13.3228	9.5754	9.6412	7.5291	11.4372	11.4734	109.20	7.9
0.50	13.3310	9.5788	9.6764	7.5373	11.4323	11.4658	109.25	8.0
1	—	—	—	7.5426	11.4391	11.4715	109.31	7.4

Therefore given gadolinium concentration from x(Gd) > 0.10 and higher in the $Sc_2Mo_3O_{12}$–$Gd_2Mo_3O_{12}$ system we can observe a coexistence of two phases with different structures: scandium molybdate and gadolinium molybdate. The structural distinctions are given in Fig. 2.6.

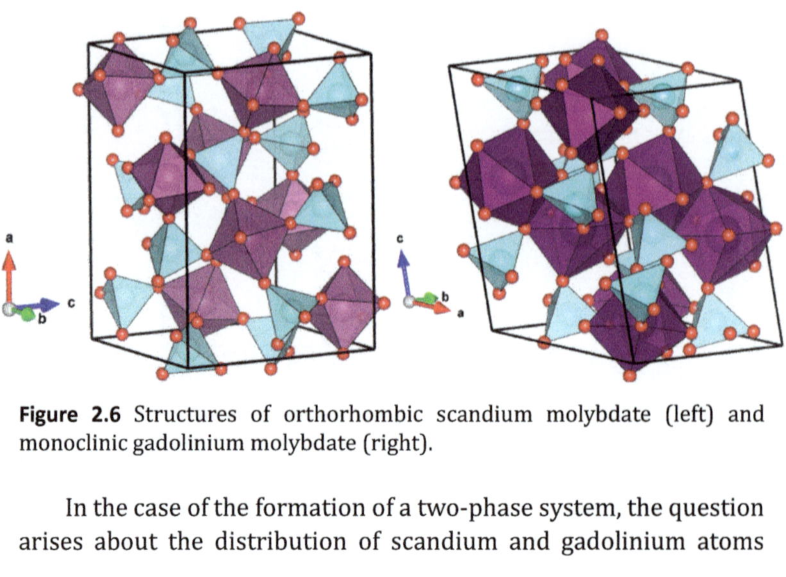

Figure 2.6 Structures of orthorhombic scandium molybdate (left) and monoclinic gadolinium molybdate (right).

In the case of the formation of a two-phase system, the question arises about the distribution of scandium and gadolinium atoms over these phases. The Rietveld method allows the occupancy of the sites by gadolinium and scandium atoms in both phases to be refined (Table 2.2).

Table 2.2 Occupancy of scandium and gadolinium atoms in solid solutions with SMO and GMO structures

	SMO phase		GMO phase	
x(Gd)	Occ. Sc	Occ. Gd	Occ. Sc	Occ. Gd
0.005	1	0	—	—
0.025	1	0	—	—
0.05	1	0	—	—
0.075	0.95	0.05	—	—
0.10	0.99	0.01	—	—
0.25	0.90	0.10	0.10	0.90
0.50	0.87	0.13	0.13	0.87
1	—	—	0	1

When gadolinium concentration is $x(\text{Gd}) < 0.075$, it appears impossible to calculate the occupancies of scandium and gadolinium due to small concentrations of the doping element. At $x(\text{Gd}) = 0.10$ the calculation shows that gadolinium atoms occupy less sites in the scandium sublattice than at $x(\text{Gd}) = 0.075$. This may indirectly testify to the formation of gadolinium molybdate phase at this concentration.

The obtained data show that given gadolinium concentration $x(\text{Gd}) > 0.10$ the system consists of two phases, which are solid solutions: the solution of gadolinium in scandium molybdate and the solution of scandium in gadolinium molybdate. This is supported by the data of occupancy of the sites by scandium and gadolinium atoms and by the changes in the unit cell parameters of both phases.

2.3 Raman Spectroscopy

In this work Raman spectroscopy was used for examination of the phases in the solid solutions with various Gd concentrations. The obtained data allow a conclusion about the only phase of scandium molybdate structure given gadolinium concentrations from 0 to 0.10. For concentration x(Gd) = 0.25 the spectrum reflects the emergence of the second phase of gadolinium molybdate. This becomes evident when correlating the spectra of pure scandium and gadolinium molybdates and the spectra of solid solutions with various gadolinium concentration (Fig. 2.7).

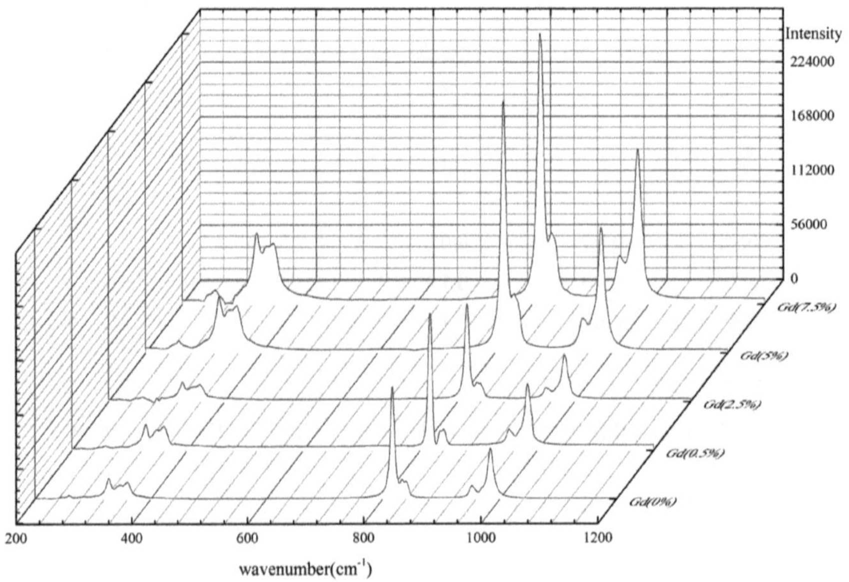

Figure 2.7 Raman spectra for the solutions with SMO structure.

The obtained spectra suggest the presence of only one phase with SMO structure in the samples. As the gadolinium concentration increases no noticeable shifts of the frequencies in the spectra are observed.

However, at gadolinium concentration x(Gd) = 0.25 a group of maxima appears in the region 900–100 cm^{-1}, but according to the relationship of the peaks intensities the phase of scandium molybdate seems to prevail, which is confirmed also by the data of quantitative X-ray analysis (Fig. 2.8).

The obtained frequencies are in good agreement with the results described in ref. 5. For the phase of scandium molybdate, the lines in the region of wavenumbers 814–817 cm^{-1} and 980–984 cm^{-1} were found (Table 2.3).

Table 2.3 Wavenumbers corresponding to the absorption bands in Raman spectra

Visual intensity	x(Gd), %									Vibration symmetry of SMO phase
	0	0.25	1.25	2.5	3.75	5	12.5	25	50	
m								303	301	A_g, B_{3g}
w	328	327	328	328	328	327	332	332 (332)	333	B_{3g}
vw	346	346	346	344	349	346		349 (349)	350	A_g, B_{1g}
vw	360	360	360	356	356	352	359	362 (362)	364	A_g, B_{2g}, B_{3g}
w	376	374	375							B_{3g}
m							396	392	390	
w								401	404	
w								441	441	
m							465	465	464	
w								595	596	
w								639	636	
w								683	680	
m							743	751	746	
m							778	785	782	
m							796	788	791	
vs	817	817	816	815	814	814	813	817		A_g, B_{1g}, B_{2g}
m	829	832	832	834	837	835	830		823	A_g, B_{3g}
vw	840	840	840				844	844	847	B_{3g}, B_{1g}
vs							902	902	903	
s							919	916	916	
vw					925		923			A_{2g}
w	954	953	958	955	954	954	954	956		B_{3g}
vw	969	971								A_g
s	984	984	983	982	981	981	980	980		A_g

Visual intensity: *vw*, very weak; *w*, weak; *m*, moderate; *s*, strong; *vs*, very strong.

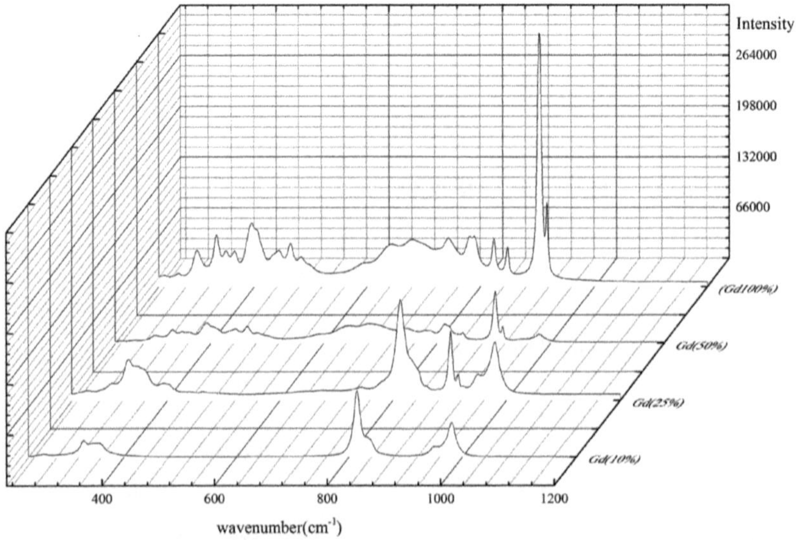

Figure 2.8 Raman spectra of the samples containing the phase with GMO structure.

The examples of the simulation of Raman spectra in the program package OriginPro 9.0 by Gauss curves are given in Figs. 2.9 and 2.10.

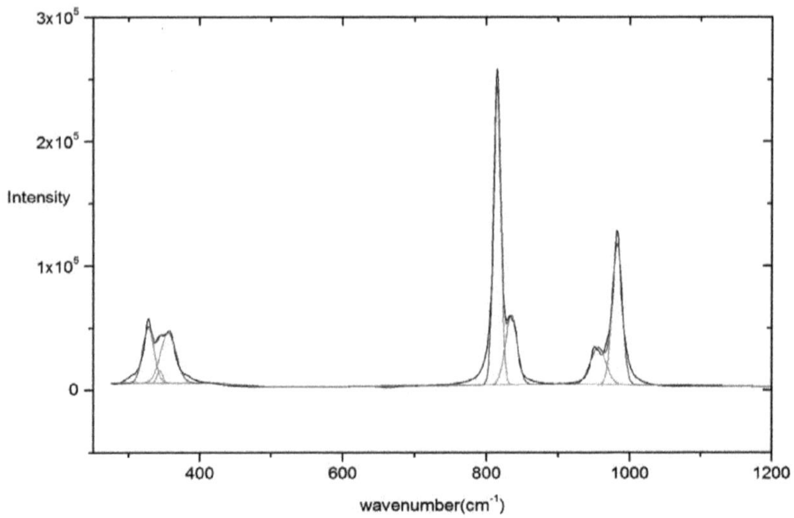

Figure 2.9 Simulation of Raman spectrum for solid solution with x(Gd) = 0.05 by Gauss curves.

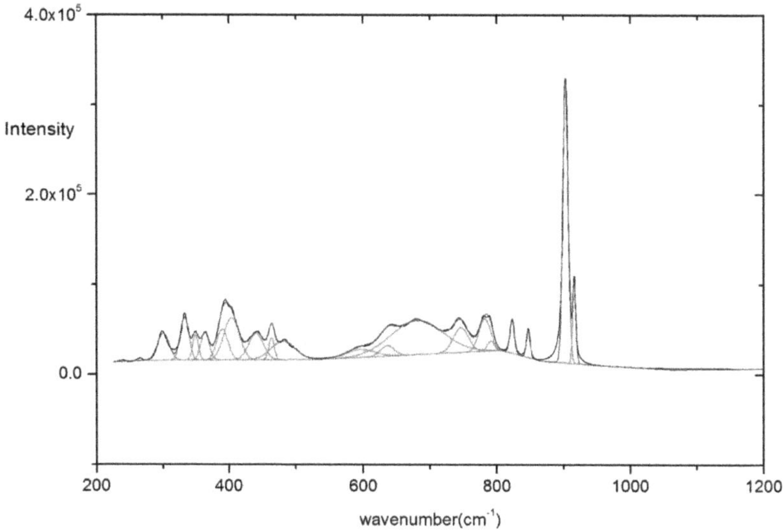

Figure 2.10 Simulation of Raman spectrum for solid solution with x(Gd) = 0.25 by Gauss curves.

2.4 Magnetic Susceptibility Measurements and Their Analysis

We also studied the magnetic susceptibility of all the solid solutions in the temperature range 77–400 K. On the basis of measurements the data on paramagnetic component and effective magnetic moment were obtained and examined.

The paramagnetic component of magnetic susceptibility obeys the Curie–Weiss law over the whole temperature range and for all the concentrations (Fig. 2.11).

In the concentration range of x(Gd) = 0.005–0.075 the paramagnetic component of magnetic susceptibility is almost constant (Fig. 2.12). This is the first evidence for the fact that in this concentration range we deal with the only phase with SMO structure, where gadolinium atoms substitute for scandium atoms. The constancy of paramagnetic component of magnetic susceptibility upon changing gadolinium concentration points to the fact that gadolinium atoms do not take part in the exchange interactions between each other, since they are located in the scandium sites and their octahedra are separated with MoO_4 tetrahedra.

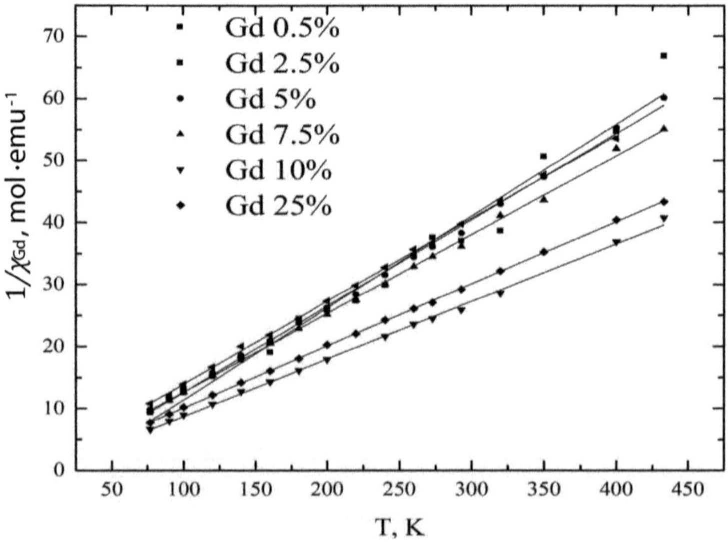

Figure 2.11 Plots of inverse paramagnetic component of magnetic susceptibility calculated per 1 mole of gadolinium atoms vs. temperature for the $Sc_{2-2x}Gd_{2x}Mo_3O_{12}$ system.

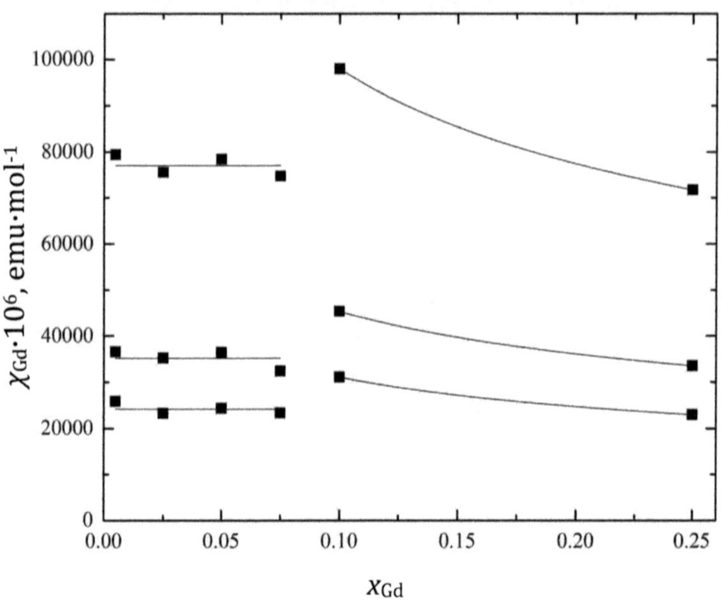

Figure 2.12 Plots of paramagnetic component of magnetic susceptibility calculated per 1 mole of gadolinium atoms vs. gadolinium concentration in the $Sc_{2-2x}Gd_{2x}Mo_3O_{12}$ solid solutions for three temperatures.

However, at x(Gd) = 0.10, a jump is observed in all the isotherms, which points to the formation of gadolinium molybdate phase with another structure differing from the structure of scandium molybdate. This results in the emergence of magnetic exchange between gadolinium atoms and, as a consequence, to an abrupt change in the magnetic susceptibility.

A decrease in the paramagnetic component of magnetic susceptibility at x(Gd) > 0.10 testifies to antiferromagnetic interactions in the system resulting in the exchange between gadolinium atoms in the GMO phase (Fig. 2.12).

As the temperature increases, the effective magnetic moment at the infinite dilution slightly decreases (Fig. 2.13). The obtained μ_{eff} are close to the effective magnetic moment for a Gd^{3+} (7.92 μB) by the data of ref. 6.

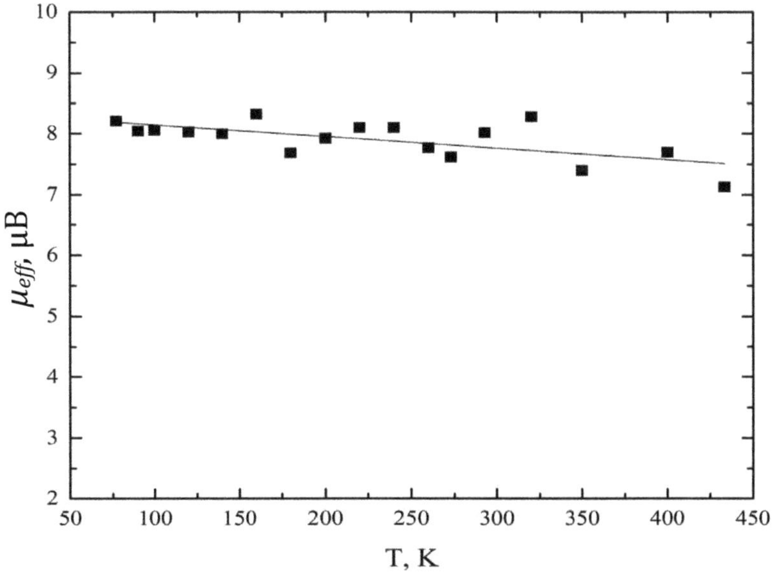

Figure 2.13 The dependence of the effective magnetic moment at the infinite dilution on temperature.

Therefore, the method of magnetic dilution and examination of the isotherms of paramagnetic components of magnetic susceptibility offer the prospect of recording and determination of the homogeneity region of solid solutions based on scandium molybdate, which is not always possible by the data of X-ray diffraction and Raman spectroscopy owing to a rather low sensitivity of these methods in the region of diluted solid solutions.

References

1. Varga T, Moats JL, Ushakov SV, Navrotsky A (2007) Thermochemistry of $A_2M_3O_{12}$ negative thermal expansion materials, *J Mater Res*, **22**(9), 2512–2521.
2. Ullman FG, Holden BJ, Ganguly BN, Hardy JR (1973) Raman spectrum of gadolinium molybdate above and below the ferroelectric transition, *Phys Rev B*, **8**(6), 2991–2998.
3. Ponomarev BK, Stiep E, Wiegelmann H, Jansen AGM, Wyder P, Red'kin BS (2000) Anisotropy of the magnetoelectric effect in β-$Gd_2(MoO_4)_3$, *Phys Sol St*, **42**(4), 734–738.
4. Shannon RD (1976) Revised effective ionic radii and systematic studies of interatomic distances in halides and chalcogenides, *Acta Cryst A*, **32**(5), 751–767.
5. Maczka M, Hermanowicz K, Hanuza J (2005) Phase transition and vibrational properties of $A_2(BO_4)_3$ compounds (A = Sc, In; B = Mo, W), *J Mol Str*, **744–747**, 283–288.
6. Kalinnikov VT, Rakitin YuV (eds) (1980) Introduction to magnetochemistry. Method of static magnetic susceptibility in *Chemistry*, Nauka, Moscow. [In Russian].

Chapter 3

Problems of Electron Structure of Colossal Magnetoresistors

Anna V. Fedorova and Natalia V. Chezhina

Department of General and Inorganic Chemistry, St. Petersburg State University, Universitetskaya nab. 7/9, St. Petersburg, 199034, Russia

natalia.chezhina@gmail.com, n.chezhina@spbu.ru

3.1 Introduction

Complex oxides with perovskite structure have been studied since the middle of the 20th century.[1,2] In spite of the fact that the first results on electrical properties of lanthanum manganites had been published more than 60 years ago,[3] the interest to them did not diminish and reached its peak when the effect of colossal magnetoresistance (CMR) was found in the oxide ceramics based on lanthanum manganite.[4] Of the greatest interest are complex oxides with the composition $La_{1-x}A_xMnO_3$ (A = Ca, Sr, Ba). The concentration of dopants (x) may vary within wide limits, the properties of ceramic samples varying substantially in doing so, which is associated with various phase transitions and types of ordering.

Electronic Structure of Materials: Challenges and Developments
Edited by Natalia V. Chezhina and Dmitry A. Korolev

ISBN 978-981-4800-55-6 (Hardcover), 978-0-429-24287-8 (eBook)
www.panstanford.com

The CMR effect consists of an abrupt decrease in the resistance upon the imposition of magnetic field. It attains the greatest values in the vicinity of Curie point[5] and $\delta_H = [\rho(H) - \rho(0)]/\rho(H)$ was found in the $La_{0.67}Ca_{0.33}MnO_3$ films. The relative resistance, δ_H, reached 1300% at room temperature.[6]

It was natural that CMR found in $LaMnO_3$ doped with calcium stimulated the interest to pure lanthanum manganite, which is the basis for obtaining magnetoresisting materials. Magnetic structure of $LaMnO_3$ by the data of neutron graphic studies[7] may be described as an antiferromagnetic lattice consisting of magnetic layers of Mn^{3+}. The study of magnetic properties showed that at low temperature $LaMnO_3$ is an antiferromagnetic and at temperatures higher than 300 K antiferromagnetic character of magnetic susceptibility defined by superexchange interactions between manganese atoms completely disappears.

It is agreed[8] that antiferromagnetic exchange decreases as the Mn–O–Mn bond angle decreases. However, we must remember that according to the exchange channel model[9] the exchange between Mn^{3+} atoms at an angle of 180° must be antiferromagnetic, whereas a decrease in this angle must result in an increase in the contribution of ferromagnetic channels of exchange. The study of the structure and magnetic properties of lanthanum manganite show that the latter depends on the content of Mn^{4+}.[10] In the range of Mn^{4+} concentration more than 14% the structures with cubic, rhombohedral, and orthorhombic structure result in ferromagnetic ordering of the spins. In this case a fraction of manganese atoms make up magnetic clusters, which is conventionally accounted for by Mn^{4+} atoms localizing near cation vacancies resulting in the *double exchange*. A decrease in Mn^{4+} concentration is characterized by antiferromagnetic ordering. At Mn^{4+} concentration 10–14 mol% ferro- and antiferromagnetic components are observed in the magnetic moment.

It is essential that ferromagnetic interactions in stoichiometric $LaMnO_3$ does not result in magnetoresistivity. However, non-stoichiometry results in some anomalies in the magnetic properties.[11,12]

To account for physical properties of manganites we must consider their crystal structure and find correlations with their electron structure.

Stoichiometric $LaMnO_3$ has a perovskite-like structure[13] (Fig. 3.1). Substitution of a fraction of lanthanum atoms for bivalent calcium atoms results in a mixed valence compound $La^{3+}_{1-x}Ca^{2+}_{x}Mn^{3+}_{1-x}Mn^{4+}_{x}O_3$. In spite of the fact that perovskite structure is reasonably tolerant to various substitutions, in reality the cubic structure occurs rather rarely. As a rule perovskite structure undergoes various distortions, which are determined either by steric specific features (a discrepancy in the sizes of dopants), or by the Jahn–Teller effect.

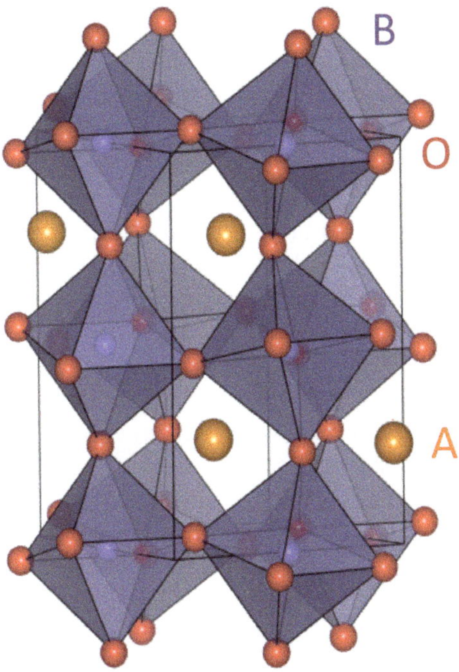

Figure 3.1 Perovskite structure.

It was found experimentally that from the point of view of CMR the best characteristics are demonstrated by $La_{1-x}A_xMnO_3$ (A = Ca, Sr, Ba) complex oxides with $x \sim 0.3$.[14–16] Given this very concentration of an alkaline earth element a metallic ferromagnetic phase appears in the doped manganites. Introduction of calcium atoms at $x > 0.5$ results in the system becoming antiferromagnetic dielectric. For a long time, lanthanum manganites doped with barium have been the less studied among all the alkaline earth manganites. This was determined by the problems of obtaining single phase samples in a wide concentration range[17,18] associated with the formation of hexagonal $BaMnO_3$ given high barium atom concentrations in $La_{1-x}Ba_xMnO_3$. Curiously, the peak of magnetoresistance was found for the composition $La_{0.5}Ba_{0.5}MnO_3$ with the temperature of the phase transition about 270 K.[19] All this leads to a conclusion about a substantial influence of the nature and concentration of diamagnetic elements on magnetic properties of lanthanum manganites.

The nature of ferromagnetic phase in manganites was accounted for by Zeener in 1951.[9] The model is based on the assumption of the existence of strong intraatomic exchange interaction between a localized spin and a delocalized electron. This mechanism was called a double exchange and assumes an electron transfer from one manganese atom to another via an intermediate oxygen atom.

In spite of experimentally determined, optimal from the point of view of CMR relationships between lanthanum and alkaline earth elements in manganites a rather large interest of the researchers is attracted to manganites with the substitution of lanthanum atoms for a half of bivalent elements ($x \sim 0.5$). This interest to materials of such a composition seems to be associated with the fact that it is at $x = 0.5$ a change from ferromagnetic to antiferromagnetic states occurs in these systems. Such a special feature of manganites given a half doping in the lanthanum sublattice made these systems promising with the aim of searching for an interrelation between spin, orbital, and charge ordering.[20,21] These studies are directed to determining the role of various factors in the stabilization of one or another type of magnetic structure.

From the applied point of view lanthanum manganites doped with strontium attract the greatest attention. They have the greatest CMR in the temperature range 40–400 K, when concentration of the substituent varies in the $0.10 \leq x \leq 0.20$ region.[22] The obtained experimental properties are accounted for within the double exchange model and are associated with the changes in the ratios between manganese atoms with various valence states. However, the problem of the mechanism of CMR effect and its association with the nature of the doping elements in the perovskite structure is still open. There is an opinion that the properties of low doped lanthanum manganites may be accounted for within the model of phase separation, its basis being constituted of the assumption about the presence of magnetic clusters in the non-conducting medium. In this case the authors postulate that the exchange energy gain and elastic stresses in the lattice result in an association of magnetic polarons of small radius into a large magnetic cluster with several charge carriers.

By now the lanthanum manganites doped with alkaline earth elements (calcium, strontium, and barium) are the most widely studied. The study of the systems with isovalent substitution of other trivalent elements for lanthanum atoms—rare earth, for example, is no less interesting. Thus introduction of yttrium atoms into the sites of lanthanum[23] showed that the changes in the electro physical properties including a decrease in the resistance in such systems as $(La_{1-}Y_x)_{0.67}(Ca/Sr)_{0.33}MnO_3$ occurs no monotonously with the variation of yttrium atom concentration. In this case the best characteristics are demonstrated by complex oxides containing 10 mol% of yttrium.

An increased interest to complex oxides based on $LaMnO_3$ revealed a large number of properties in the doped manganites, which required physical and chemical explanation. In spite of a large number of models used nowadays for explaining electrophysical characteristics, CMR among them, the microscopic nature of this effect remains the subject of acute discussions. Since the properties of magnetoresistive ceramics essentially depend on the composition of complex oxide, of special importance is the problem of obtaining materials with predetermined CMR by selecting the doping elements and their ratios. Just this problem has not yet been solved. The estimation of the influence of doping elements on magnetic and electrical characteristics is a very complicated challenge. By common opinion the effect of CMR is cooperative and determined by the exchange interactions between paramagnetic centers. Oxide ceramics is usually magnetically concentrated. Their study is complicated by a large number of electron and other cooperative effects. We see the best way to solve this problem in using the magnetic dilution method, which successfully demonstrated its feasibility, as shown in Chapter 1. We bear in mind that diluted solid solutions of lanthanum manganites in diamagnetic matrices will not show CMR effect in themselves, but the method can reveal the variations in the short order exchange interactions and clustering in the concentrated manganites, thus allowing substantial correlations between the nature and concentration of doping elements and electro physical properties to be derived. The main approach to the problem is that we first vary the nature and concentration of only one component of the system.

3.2 State of Manganese Atoms and Exchange Interactions in the Solid Solutions Based on Lanthanum Aluminate

3.2.1 The $xLa_{0.33}A_{0.67}MnO_3$–$(1–x)LaAlO_3$ (A = Ca, Sr, Ba) Solid Solutions

With the aim of revealing the influence of the nature of alkaline earth elements on the state of manganese atoms and the character of exchange interactions between them, we studied the $xLa_{0.33}A_{0.67}MnO_3$–$(1–x)LaAlO_3$ (A = Ca, Sr, Ba) solid solutions by the method of magnetic dilution in the concentration range $x = 0.01–0.10$.

For all the solid solutions under study obtained by ceramic procedure, we measured the magnetic susceptibility in the temperature range 77–400 K, calculated the paramagnetic components of magnetic susceptibility related to 1 mole of manganese atoms and the effective magnetic moments. Concentration and temperature dependencies of magnetic characteristics were plotted.

A comparison of magnetic properties of solid solutions containing various alkaline earth elements shows the following regularities. First of all the isotherms of paramagnetic component of magnetic susceptibility are drastically different for the solid solutions containing calcium, strontium, and barium (Fig. 3.2). Taking into account the fact that in the diluted solid solutions the fraction of bivalent element is extremely small ($0.33x$, given x no more than 0.1), we must assume that the formation of clusters of manganese atoms occurs in the immediate vicinity and consequently with participation of alkaline earth element.

An increase in the paramagnetic component of magnetic susceptibility χ_{Mn} with manganese atom concentration and the temperature changes in the magnetic moment in the case of $x\text{La}_{0.33}\text{Ca}_{0.67}\text{MnO}_3$–$(1-x)\text{LaAlO}_3$ suggested that in the solid solutions with $x > 0.04$, the paramagnetic atoms are bonded in clusters containing more than two manganese atoms.[24] At low temperatures the exchange interactions in these clusters are ferromagnetic ($J > 0$), no matter which atoms take part in the exchange—with different spins or with the same spin. In this case the exchange parameter varies as the temperature increases and becomes antiferromagnetic ($J < 0$). This is possible if we assume that at low temperatures there are local distortions in the Mn^{3+}–O–Mn^{3+} valence angles, which disappear due to heat motion as the temperature increases. In such a case ferromagnetic exchange between atoms with the same spin gives way to antiferromagnetic in agreement with exchange channel model.[9]

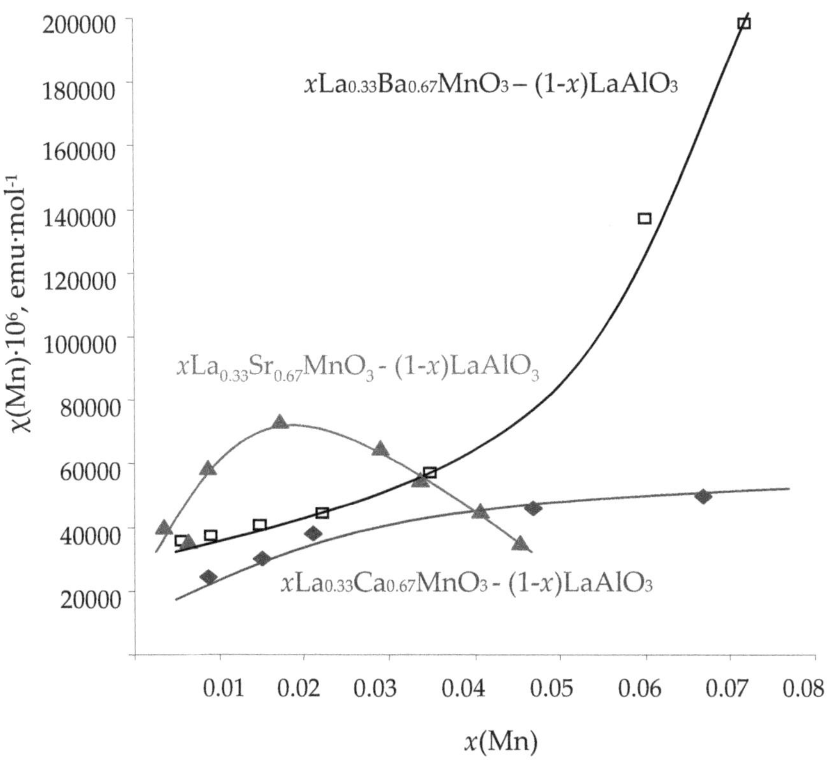

Figure 3.2 Plot of paramagnetic component of magnetic susceptibility, χ_{Mn} vs. manganese concentration at 100 K for various systems.

Introduction of strontium atoms into the system changes the pattern essentially. The $xLa_{0.33}Sr_{0.67}MnO_3$–$(1–x)LaAlO_3$ system[25] is distinguished by a maximum in the isotherms of magnetic susceptibility in the range of paramagnetic atom concentration $x \sim 0.02$. This points to the formation of clusters with ferromagnetic exchange, but further increase in x results in greater aggregation of paramagnetic atoms accompanied by the formation of larger aggregates with antiferromagnetic component of the exchange prevailing. This may be due to greater size of the space of clusters owing to the participation of larger strontium cations, thus less pronounced thermal expansion.

In the case of lanthanum manganites doped with barium, the extrema are absent,[26] the paramagnetic components of magnetic susceptibility monotonously increase as the concentration increases. Extrapolation of paramagnetic component of magnetic susceptibility and of effective magnetic moment to the infinite dilution of the $xLa_{0.33}Ba_{0.67}MnO_3$–$(1–x)LaAlO_3$ solid solutions ($x \to 0$) showed that even at the infinite dilution ferromagnetic clusters are preserved.

Therefore, it can be said with confidence that the nature of doping alkaline earth element substantially influences the clustering and exchange interactions in lanthanum manganites.

When considering the physical properties of substituted perovskite like manganites, the difference in the radii of substituting atoms appears to be one of important criteria. In spite of a tolerance of perovskite structure to various substitutions, in complex oxide systems distinct distortions are conventionally observed. When calcium atoms close in their radius to lanthanum ($r(La^{3+})_{XII}$ = 1.35 Å and $r(Ca^{2+})_{XII}$ = 1.32 Å) by Shannon[27] for coordination number 12 substitute for La, the structure must not undergo rigid distortions. Passing from calcium to strontium and barium with larger sizes, the local distortions of the structure occur, and they become more rigid. An increase in the sizes of doping atoms in its turn results in the fact that the distortion of the Mn–O–Mn bond angles becomes more stable with temperature.

Along with the size effect, the degree of bond ionicity is no less important. Upon passing from calcium to strontium and barium, the ionicity of the bond between alkaline earth atom and oxygen increases. The covalence of the Mn–O bond increases simultaneously. Since antiferromagnetic exchange is associated with orbital overlapping, the more covalent is Mn–O bond, the stronger is antiferromagnetic exchange. The more so aggregation of manganese atoms also increases as the covalency of Mn–O bond increases.[28] Just this is observed in the case of xLa$_{0.33}$Ba$_{0.67}$MnO$_3$–(1–x)LaAlO$_3$ solid solutions.

3.2.2 The xLa$_{0.67}$(Ca$_y$Sr$_{1-y}$)$_{0.33}$MnO$_3$–(1–x)LaAlO$_3$ (y = 0.3; 0.5; 0.7) Solid Solutions

The search for the composition of the materials with predetermined properties is often carried out by simultaneous introduction of various alkaline elements into the perovskite structure.[29,30]

We studied the xLa$_{0.67}$(Ca$_y$Sr$_{1-y}$)$_{0.33}$MnO$_3$–(1–x)LaAlO$_3$ solid solutions with the aim to reveal the influence of simultaneous doping the perovskite structure with calcium and strontium on the state of manganese atoms and the character of exchange interactions in manganites. Various relationships between calcium and strontium atoms were studied (y = 0.3; 0.5; 0.7).[31] For all the solid solutions obtained by ceramic procedure, the paramagnetic components of magnetic susceptibility were calculated per 1 mole of manganese atoms and the effective magnetic moments. The isotherms of magnetic susceptibility are similar in shape for all the solid solutions—χ_{Mn} monotonously increases as the concentration of Mn increases (Fig. 3.3), which points to a domination of ferromagnetic component in the exchange.

We can see that a partial change of calcium for strontium substantially changes the run of concentration dependence of magnetic susceptibility in comparison to the La$_{0.67}$Ca$_{0.33}$MnO$_3$ and La$_{0.67}$Sr$_{0.33}$MnO$_3$ in LaAlO$_3$ solid solutions. This can be accounted for by an intensification of ferromagnetic exchange to a greater extent than is observed in the systems containing only calcium or only strontium. For all the solid solutions containing both bivalent elements, magnetic susceptibility almost does not depend on paramagnetic content in a narrow concentration range ($0 < x < 0.02$); however, further increase in concentration results in an abrupt increase in magnetic susceptibility, which points to an enhancement of ferromagnetic exchange.

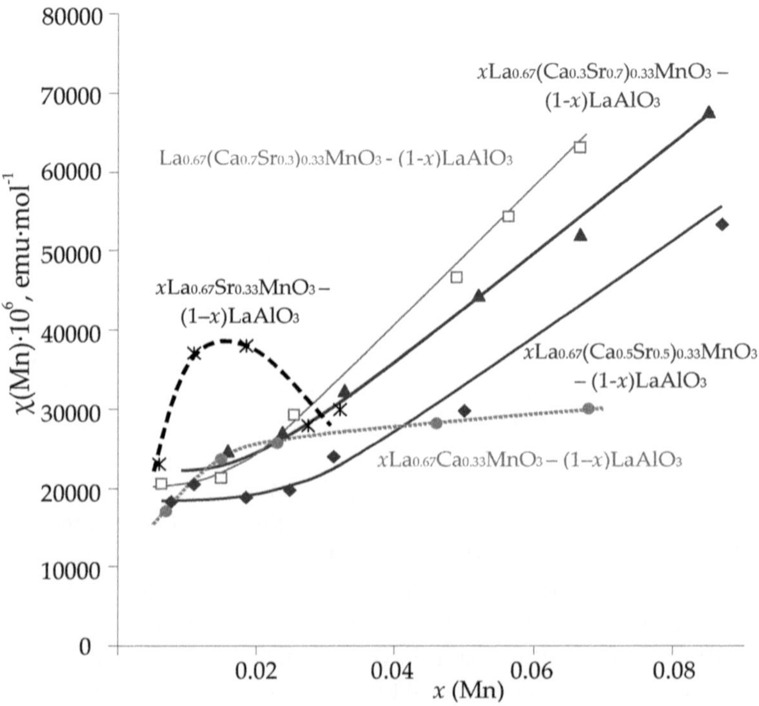

Figure 3.3 Plots of paramagnetic component of magnetic susceptibility vs. manganese concentration for various systems.

It is interesting that the location of χ_{Mn} isotherms does not obey any regularity. For all the solid solutions under study, magnetic susceptibility reaches the greater values at the Ca:Sr ratio 0.7:0.3, and the smallest at Ca:Sr = 0.5:0.5. In the last case the isotherms of magnetic susceptibility at small concentrations are even lower than for manganites doped only with calcium. Temperature dependencies of the effective magnetic moment also testify to the same assumption. In this case the temperature dependencies of the effective magnetic moments also are similar for all three systems with various relationships between doping elements.

Given a small manganese content, μ_{eff} almost does not depend on temperature and upon extrapolation to the infinite dilution can be described as the sum of magnetic moments of single Mn^{3+} and Mn^{4+} atoms like that in Fig. 3.4 ($\mu_{S.O.}(Mn^{3+})$ = 4.90 μB, $\mu_{S.O.}(Mn^{4+})$ = 3.88 μB). However, as the concentration of Mn increases, the effective magnetic moment decreases with temperature to a growing extent. The data of extrapolation of magnetic characteristics to the infinite dilution point to the fact that at $x \to 0$ a complete disaggregation of manganese atoms occurs.

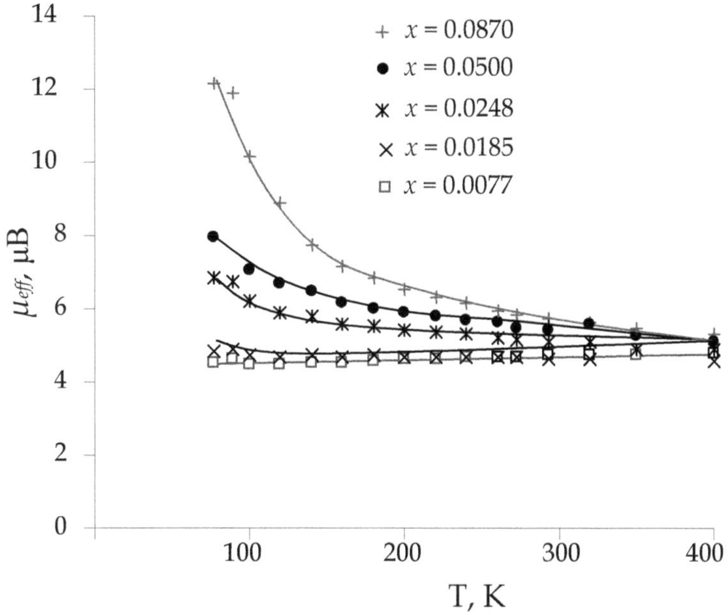

Figure 3.4 Plots of effective magnetic moments vs. temperature for the $x\mathrm{La_{0.67}(Ca_{0.5}Sr_{0.5})_{0.33}MnO_3}$–$(1-x)\mathrm{LaAlO_3}$ system with various concentrations of manganese (x).

Table 3.1 μ_{eff} and the fraction of Mn^{3+} at the infinite dilution in the $x\mathrm{La_{0.67}(Ca_ySr_{1-y})_{0.33}MnO_3}$–$(1-x)\mathrm{LaAlO_3}$ solid solutions

Ca:Sr	μ_{eff} (μB)	a
0.3:0.7	4.6	0.68
0.5:0.5	4.5	0.58
0.7:0.3	4.4	0.48

On the basis of experimental effective magnetic moments for the $x\mathrm{La_{0.67}(Ca_ySr_{1-y})_{0.33}MnO_3}$–$(1-x)\mathrm{LaAlO_3}$ solid solutions at the infinite dilution, the fraction of Mn^{3+} atoms was calculated by formula

$$\mu_{\text{eff}}^2(\text{Mn}) = a\mu_{\text{Mn}^{3+}}^2 + (1-a)\mu_{\text{Mn}^{4+}}^2,$$

where a is the fraction of Mn^{3+}.

First of all the fact that fixes attention is that in all the solid solutions doped with calcium and strontium simultaneously, the ratio Mn^{3+}:Mn^{4+} = 0.67:0.33 is not fulfilled. This may be due to comparatively large size of Mn^{3+} atoms and their ability to be oxidized to Mn^{4+} at the expense of a certain non-stoichiometry in the sublattice of heavy metals.

According to our calculations as the concentration of strontium atoms in $x\text{La}_{0.67}(\text{Ca}_y\text{Sr}_{1-y})_{0.33}\text{MnO}_3$–$(1-x)\text{LaAlO}_3$ increases, the fraction of Mn^{3+} atoms increases. A similar tendency is also observed in the solid solutions containing only calcium or only strontium, which can be accounted for by the fact that strontium has an essentially greater radius than calcium ($r(\text{Sr}^{2+})$ = 1.44 Å), which results in an increase in the unit cell size, in an enlargement of octahedral spaces occupied by manganese atoms.

A minor increase in the paramagnetic component of magnetic susceptibility at low manganese atom concentration (x) suggests that in the concentration range $x < 0.02$ two types of dimers are formed: Mn^{3+}–O–Mn^{4+} and Mn^{3+}–O–Mn^{3+} determining ferro- and antiferromagnetic exchange, respectively. Further increase in the manganese fraction results in an enlargement of paramagnetic clusters, perhaps to tetramers and greater, and to an increase in the ferromagnetic component in the exchange.

Therefore, in the case of the ratio Ca:Sr = 0.7:0.3, 0.5:0.5, and to a lesser degree 0.3:0.7 in the diluted solid solutions first of all clusters of two manganese atoms are formed and the competition of antiferromagnetic and ferromagnetic exchange in them results in almost constant μ_{eff} and χ_{Mn} in the concentration range $0 < x < 0.02$.

In this case it appears that given small content of paramagnetic element in the system with the ratio Ca:Sr = 0.3:0.7 the effective magnetic moment attains ~6 μB and has a tendency to a small decrease, which points to a predominance of ferromagnetic exchange. Whereas in the solid solutions with Ca:Sr = 0.5:0.5 μ_{eff} does not exceed 5 μB in the same concentration range. This testifies to an almost complete compensation of antiferromagnetic exchange in the dimers with the same oxidation state and ferromagnetic exchange in the dimers Mn^{3+}–O–Mn^{4+}.

3.2.3 The $x(La_{1-z}Y_z)_{0.67}Ca_{0.33}MnO_3$–$(1-x)La_{1-z}Y_zAlO_3$ (z = 0.1; 0.2) Solid Solutions

Recent works on magnetoresistance are devoted to the influence of rare earth elements introduced into the sublattice of lanthanum on the CMR effect.[32-35] Since most rare earth elements are paramagnetic, and thus can take part in the exchange first between themselves and second with manganese atoms, we concentrated our attention on the influence of yttrium atoms on the states of manganese atoms and exchange interactions in the lanthanum manganites with partial substitution of lanthanum for yttrium.

We studied the $x(La_{1-z}Y_z)_{0.67}Ca_{0.33}MnO_3$–$(1-x)La_{1-z}Y_zAlO_3$ solid solutions by the magnetic dilution method taking various contents of yttrium atoms both in the paramagnetic system and in the diamagnetic matrix (z = 0.1; 0.2).

As was discussed earlier, in the $xLa_{0.33}Ca_{0.67}MnO_3$–$(1-x)LaAlO_3$ solid solutions containing no yttrium paramagnetic component of magnetic susceptibility monotonously increased over the whole range of manganese concentrations testifying to the presence of ferromagnetically bonded clusters of manganese atoms.

Introduction of z(Y) = 0.1 into lanthanum sites resulted in essential changes in the magnetic properties of the solid solutions (Fig. 3.5). The paramagnetic components calculated per 1 mole of manganese atoms monotonously decrease in the region of small concentrations of paramagnetic atoms pointing to antiferromagnetic interactions between paramagnetic centers. As the fraction of manganese exceeds $x \sim 0.04$, the paramagnetic susceptibility begins to increase, which points to a domination of ferromagnetic clusters of paramagnetic atoms.

The dependence of paramagnetic susceptibility on manganese atoms content for the solid solutions with z(Y) = 0.2 is given in Fig. 3.6. The run of the isotherms shows substantial dissimilarity both from the solid solutions containing z(Y) = 0.1 and from lanthanum manganites doped only with calcium.

In this case, in the most diluted solid solutions ($x < 0.02$), a certain increase in the paramagnetic component of magnetic susceptibility is observed, which points to ferromagnetic exchange within paramagnetic clusters. In the concentration region $x \sim 0.02$, the magnetic susceptibility is maximal and then decreases.

At $x > 0.06$ χ_{Mn} again increases; however, at high temperature this increase becomes less pronounced.

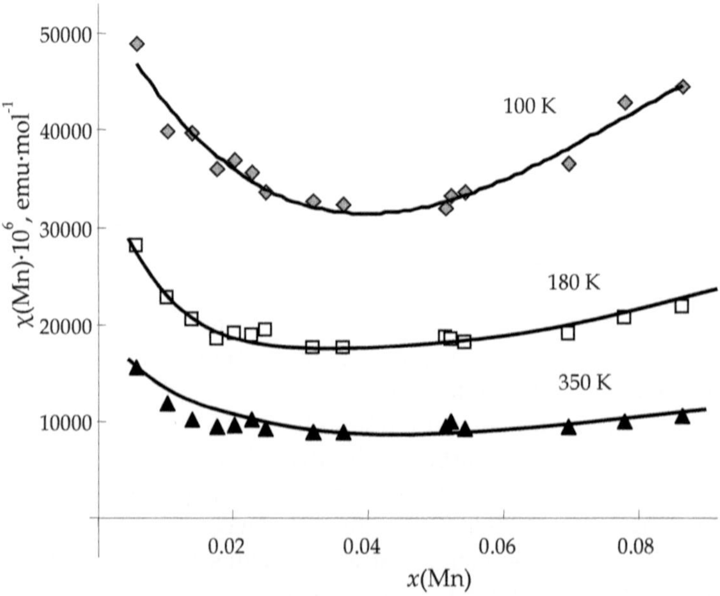

Figure 3.5 Plot of paramagnetic component of magnetic susceptibility vs. manganese concentration for the $x(La_{0.9}Y_{0.1})_{0.67}Ca_{0.33}MnO_3$–$(1\text{–}x)La_{0.9}Y_{0.1}AlO_3$ solid solutions.

A comparison between χ_{Mn} isotherms for the systems with various fractions of yttrium atoms and lanthanum manganite doped only with calcium shows the following. As is seen in Fig. 3.7, the isotherms of paramagnetic components of magnetic susceptibility differ both by the shape of the curves and by the values of χ_{Mn}.

The difference between the isotherms of magnetic susceptibility for yttrium containing manganites is the most pronounced in the region of large dilution at $x \leq 0.03$. This points to the fact that the influence of various quantities of yttrium manifests itself within short order—in the relatively small aggregates of paramagnetic atoms. An increase in the fraction of paramagnetic element levels this difference, and in the region of $x > 0.03$ the run of the isotherms has the same character independently on the content of doping yttrium atoms.

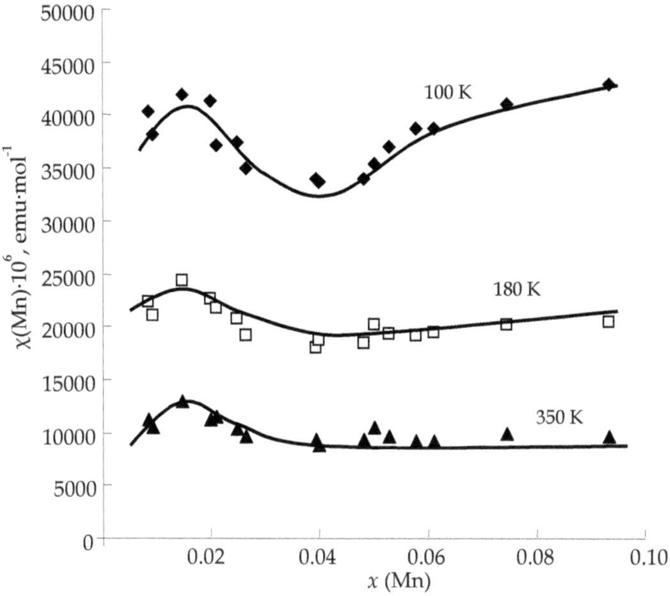

Figure 3.6 Plot of paramagnetic component of magnetic susceptibility vs. manganese concentration for the $x(La_{0.8}Y_{0.2})_{0.67}Ca_{0.33}MnO_3$–$(1–x)La_{0.8}Y_{0.2}AlO_3$ solid solutions for three temperatures.

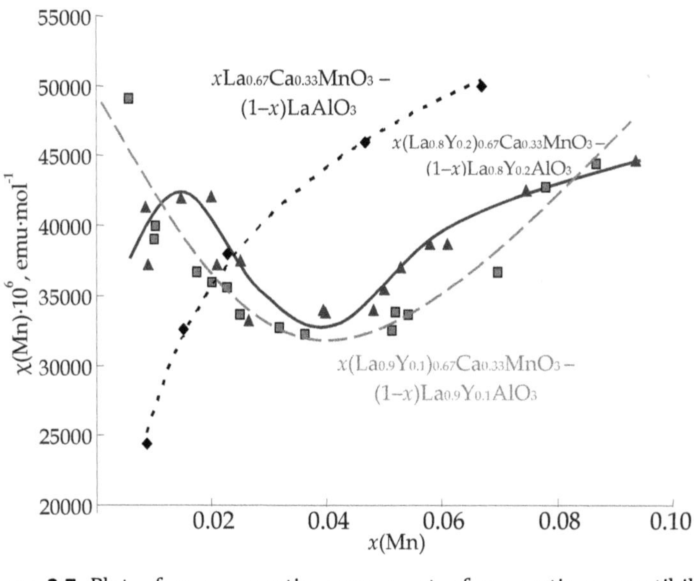

Figure 3.7 Plot of paramagnetic component of magnetic susceptibility vs. manganese concentration at 100 K for various Ca-containing systems.

A partial change of yttrium atoms for lanthanum results in essential and ambiguous changes in magnetic characteristics of these systems depending on the depth of doping. The shape of isotherms of magnetic susceptibility is rather complex in these cases and points to a non-monotonous dependence of magnetic properties of the solid solutions under study in accordance with the composition of diamagnetic sublattice.

The effective magnetic moments at the infinite dilution of the solid solutions obtained by extrapolating χ_{Mn} to $x \to 0$ over the whole range of temperatures are given in Table 3.2.

Table 3.2 Effective magnetic moments at the infinite dilution for the $x(La_{1-z}Y_z)_{0.67}Ca_{0.33}MnO_3$–$(1-x)La_{1-z}Y_zAlO_3$ solid solutions as function of temperature

T, K		**77**	**90**	**100**	**120**	**160**	**180**	**220**	**260**	**293**	**320**	**350**
$\mu_{eff}^{x \to 0}$, μB	$z = 0.1$	6.35	6.49	6.54	6.60	6.85	6.94	7.05	6.91	6.64	7.00	6.93
	$z = 0.2$	5.11	4.97	5.22	5.11	5.08	4.88	4.87	5.05	5.18	4.92	5.04

For the $xLa_{0.33}Ca_{0.67}MnO_3$–$(1-x)LaAlO_3$ solid solutions the effective magnetic moment at the infinite dilution is 4.1 μB and does not depend on temperature. This tendency points to the fact that at $x \to 0$ only single manganese atoms are present in the system–Mn^{3+} and Mn^{4+}. Their spin only μ_{eff} are 4.90 and 3.88 μB, respectively.

Introduction of yttrium into the heavy metal sublattice is followed by an increase in μ_{eff} ($x \to 0$) in comparison to Ca-doped manganites. Moreover, the effective magnetic moment depends on temperature, and cannot be described in the context of single paramagnetic atoms. Therefore, in the solid solutions of calcium substituted manganites containing 10 and 20 mol% of yttrium no complete disaggregation of manganese atoms occurs, and even at the infinite dilution some clusters of paramagnetic atoms are present.

As can be seen from temperature dependencies of the effective magnetic moments at the infinite dilution (Fig. 3.8), for the systems with various contents of yttrium, the magnetic moment for the solid solutions with 10 mol% of yttrium μ_{eff} attains the greatest values (6.5–7.0 μB) and moderately increases with temperature.

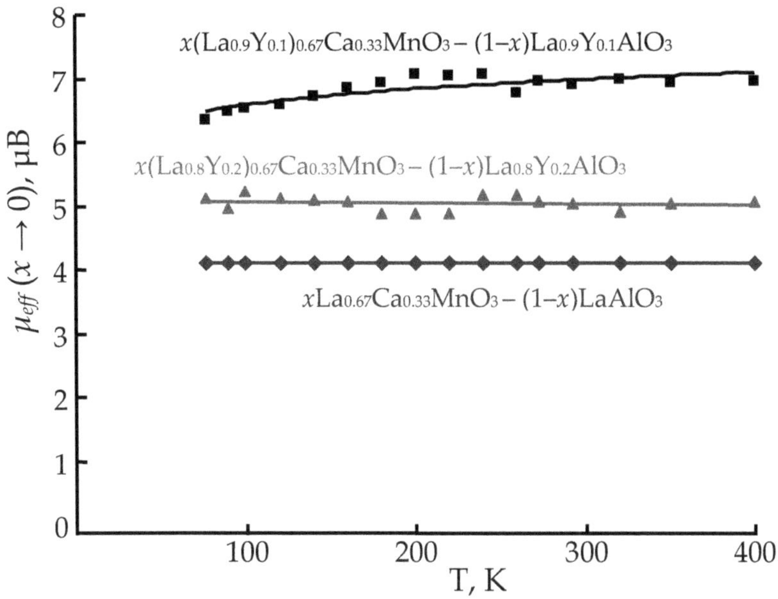

Figure 3.8 Temperature dependences of the effective magnetic moments at the infinite dilution for various systems.

Such a large effective magnetic moments cannot be accounted for by single manganese atoms Mn^{3+} (term ^5E_g), the effective magnetic moment being 4.90 μB or Mn^{4+} (term $^4A_{2g}$) μ_{eff} = 3.88 μB. The possibility of manganese reduction to Mn^{2+} (term $^6A_{1g}$) may be excluded in our case since the solid solutions were sintered in air, and sintering in the flow of oxygen, which was to oxidize manganese was proved[28] to preserve the values of magnetic susceptibility.

Since all the possible valence states of manganese atoms have orbitally non-degenerate (*A*) or doubly degenerate ground states (*E*), the magnetic moments of clusters of various sizes can be considered in the context of isotropic exchange model of Heisenberg–Dirac–Van Vleck[32] (see Chapter 1).

If we assume that at the infinite dilution the dimer clusters of manganese atoms with ferromagnetic type of the exchange are preserved in the solid solutions, the magnetic moment given large positive exchange parameter *J* may be the following within Mn^{3+}–O–Mn^{3+} 6.32 μB, Mn^{3+}–O–Mn^{4+} 5.61 μB, and for Mn^{4+}–O–Mn^{3+} 4.90 μB.

However, neither of such situations is realized in the system under study, where μ_{eff} ($x \to 0$) = 6.5–7.0 μB. It is evident that the clusters are larger. The composition of trimer clusters and corresponding μ_{eff} ($x \to 0$) may be the following: Mn^{3+}–O–Mn^{3+}–O–Mn^{3+} 7.50 μB; Mn^{3+}–O–Mn^{4+}–O–Mn^{3+} 6.90 μB; Mn^{3+}–O–Mn^{4+}–O–Mn^{4+} 6.30 μB; Mn^{4+}–O–Mn^{4+}–O–Mn^{4+} 5.74 μB.

Therefore, the effective magnetic moment ~7 μB in the solid solutions with 10 mol% content of yttrium may result from formation of trimer clusters of manganese atoms with different oxidation states.

We could also assume the presence of ferromagnetic tetramers consisting of two Mn^{3+} and two Mn^{4+} atoms (the exchange between atoms with the same oxidation state must be antiferromagnetic). However, such a cluster must have the magnetic moment ~7.9 μB and must not depend on temperature given large *J*. However, if tetramers contain three Mn^{3+} and one Mn^{4+} atom (which agrees with the ratio Mn^{3+}:Mn^{4+} in doped lanthanum manganites), the exchange between Mn^{3+} atoms being antiferromagnetic can result in a moderate increase in the effective magnetic moment with temperature.

Therefore, we may suggest that in the solid solutions containing 10 mol% of yttrium, the trimer clusters either interact with each other giving rise to antiferromagnetic exchange and a decrease in the magnetic susceptibility with concentration, or, which is more probable at higher concentrations, enlarge to tetramers consisting of three Mn^{3+} and one Mn^{4+} atoms—$[Mn_3^{3+}Mn^{4+}YCaO_{16}]^{14-}$. Further increase in the size and number of clusters results in an increase in the ferromagnetic component in the exchange.

In lanthanum manganites with 20 mol% of yttrium the magnetic moment at $x \to 0$ decreases to 5.0 μB. It is evident that in this case the disaggregation upon dilution is stronger than in the case of 10 mol% of yttrium doping, but at the infinite dilution we also cannot suggest the presence of only single paramagnetic atoms. The magnetic moment 5 μB seems to reflect the formation of Mn^{3+}–O–Mn^{4+} dimers of the composition $[Mn^{3+}Mn^{4+}YCa_{11}]^{10-}$ with ferromagnetic exchange and a rather large admixture of monomers. As x(Mn) increases, the clusters with ferromagnetic exchange are enlarged, and the effective magnetic moment increases to 5.9 μB, and only at x ~ 0.2 the formation of greater aggregates with antiferromagnetic exchange begins like in the case of $z = 0.1$. At a certain concentration of paramagnetic, the magnetic susceptibility begins to decrease. Extrapolation of magnetic susceptibility to the infinite dilution from the concentration region $x > 0.02$ (the part of the dependence after the maximum describing antiferromagnetic exchange) results in $\mu_{eff} \sim 6.51$ μB, which corresponds to the formation of clusters with the same composition as in the case of solid solutions with $z = 0.1$.

Therefore, in distinction to the solid solutions of lanthanum manganites doped only with calcium introduction of yttrium into the sites of lanthanum results in greater clustering of manganese atoms. However, the paramagnetic susceptibilities and effective magnetic moments at the infinite dilution point to a non-monotonous character of the changes in the composition of clusters—the solid solutions with 20 mol% of yttrium show a greater disaggregation of paramagnetic atoms. Introduction of 10 mol% of yttrium results in the formation of greater and more rigid clusters of manganese atoms including both calcium and yttrium atoms into their composition.

At low manganese concentrations in the systems with 10 mol% of yttrium substitution, an increase in μ_{eff} with temperature is observed. Since at the infinite dilution in the solid solution rather large clusters of paramagnetic manganese atoms with predominantly ferromagnetic exchange are preserved, an increase in μ_{eff} with temperature points to a further enlargement of clusters, but the exchange in them becomes predominantly antiferromagnetic. Only after the minimum in the concentration dependence of paramagnetic susceptibility ($x \sim 0.04$) (when the susceptibility begins to increase with increasing x), we observe an abrupt decrease in the effective magnetic moment testifying to ferromagnetic exchange. However, we must emphasize that μ_{eff} at 77 K does not attain such large values as in the case of samples containing only calcium, though an abrupt decrease in μ_{eff} points to a variable exchange parameter in large clusters.

In the solid solutions containing 20 mol% of yttrium the run of the temperature dependencies of the effective magnetic moments confirms the presence of paramagnetic clusters with two types of exchange interactions in the system.

In the region of diluted solid solutions (x < 0.02) at low temperatures, the effective magnetic moment moderately decreases, which is defined by the clusters with ferromagnetic exchange. Further increase in the fraction of manganese atoms results in an enlargement of paramagnetic aggregates, the character of exchange interactions within them changing, and μ_{eff} increases with temperature. The solid solutions with x > 0.04 show the opposite tendency—increasingly more abrupt decrease in μ_{eff} with temperature testifies to an increase in the aggregation of paramagnetic centers with ferromagnetic component prevailing in the exchange.

The temperature dependencies of the effective magnetic moment for the systems with different content of yttrium and for the $x La_{0.67}Ca_{0.33}MnO_3$–$(1-x)LaAlO_3$ solid solutions show considerable variations (Fig. 3.9).

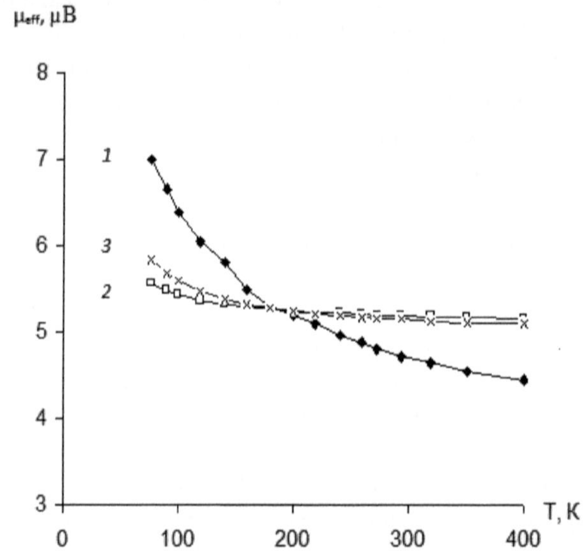

Figure 3.9 Plots of effective magnetic moments vs. temperature for the systems: (1) $x La_{0.33}Ca_{0.67}MnO_3$–$(1-x)LaAlO_3$ (x = 0.0668); (2) $x(La_{0.9}Y_{0.1})_{0.67}Ca_{0.33}MnO_3$–$(1-x)La_{0.9}Y_{0.1}AlO_3$ (x = 0.0543); (3) $x(La_{0.8}Y_{0.2})_{0.67}Ca_{0.33}MnO_3$–$(1-x)La_{0.8}Y_{0.2}AlO_3$ (x = 0.0579) with close manganese atom content.

First, the dependencies of μ_{eff} for yttrium–lanthanum solid solutions are less pronounced and in the region of sufficiently large manganese concentrations, where in both cases ferromagnetic exchange prevails, they only moderately decrease with temperature. In the solid solutions containing no yttrium the effective magnetic moment decreases from 7 to about 4.5 μB, as the temperature increases, which was accounted for by the presence of clusters with the exchange parameter J depending on temperature. First of all, we emphasize that a decrease in μ_{eff} with increasing temperature is conventionally associated with ferromagnetic exchange parameter ($kT/J > 0$). In this case the effective magnetic moment for the dimers with spins 3/2–3/2; 2–2; 3/2–2 does not exceed 6.32 μB, for a square of four atoms it does not exceed 8.5 μB. And the decrease in the temperature under study must not exceed 3%.[28] In the solid solutions studied in our work, this difference reached 35%. Therefore, even if all the manganese atoms are bonded in ferromagnetic clusters, it is impossible to describe the exchange interactions with the help of one, let it be an averaged exchange parameter.

Figure 3.10 shows the dependencies of the effective magnetic moment and J on temperature calculated with the help of Heisenberg–Dirac–Van Vleck model for the interactions between two atom with $S = 3/2$ and 2.

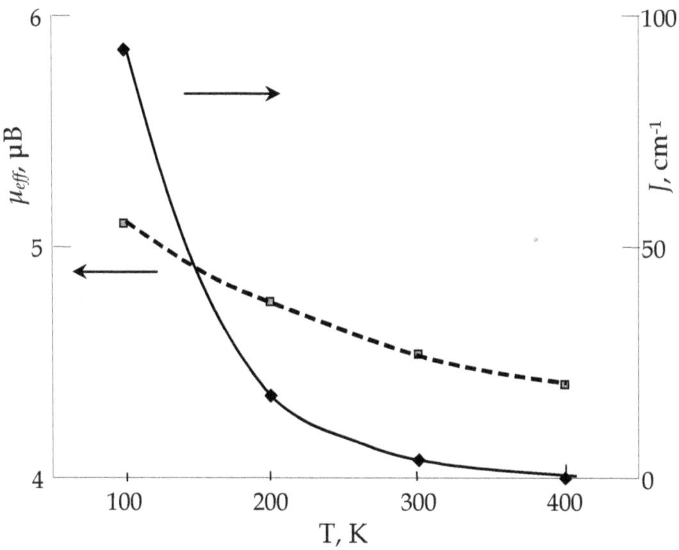

Figure 3.10 Theoretical dependencies of μ_{eff} and J on temperature for dimer clusters with $S_1 = 3/2$, $S_2 = 2$.

In the temperature range under study (77–400 K) the change in μ_{eff} from the possible maximum at $J > 0$ (5.61 μB) to the minimum (4.43 μB) corresponds to an abrupt decrease in the exchange parameter from $J \geq 90$ cm^{-1} to zero. In the case of the system under study such a conclusion is quite adequate and is associated with the fact that at low temperatures in the perovskite structure there are distortions of the Mn–O–Mn angles, resulting in ferromagnetic exchange even between Mn^{3+}–O–Mn^{3+} atoms, which must be antiferromagnetic under the angle 180°.[28] An increase in temperature results in the expansion of the lattice and vibrations of the atoms, which removes the angle distortions, and the exchange parameter decreases to almost zero at 400 K. This means that at 400 K the ferromagnetic exchange between Mn^{3+}–O–Mn^{4+} is balanced by antiferromagnetic exchange Mn^{3+}–O–Mn^{3+}.

We can also conclude that the introduction of 10 mol% of yttrium results in a stronger interactions of paramagnetic centers compared to the solid solutions containing larger quantities of the doping element (Y). In the last case the aggregation of magnetic ions appears to be weaker than in the case of $z = 0.1$.

3.2.4 The $x(La_{1-z}Y_z)_{0.67}Sr_{0.33}MnO_3$–$(1-x)La_{1-z}Y_zAlO_3$ ($z = 0.1; 0.2$) Solid Solutions

With the aim of revealing correlations between the changes in the composition of the solid solutions and their electron structure, we studied magnetic characteristics of the $x(La_{1-z}Y_z)_{0.67}Sr_{0.33}MnO_3$–$(1-x)La_{1-z}Y_zAlO_3$ solid solutions ($z = 0.1; 0.2$).

Figure 3.11 shows the concentration dependencies of paramagnetic component of magnetic susceptibility, χ_{Mn}–x, for strontium–yttrium manganites $x(La_{1-z}Y_z)_{0.67}Sr_{0.33}MnO_3$–$(1-x)La_{1-z}Y_zAlO_3$ for the case of $z = 0.1$.

As is seen from Fig. 3.11 in the region of small concentration of paramagnetic element χ_{Mn} almost does not depend on x. Further increase in manganese atoms concentration results in an essential increase in χ_{Mn} and at $x > 0.03$ clusters with ferromagnetic exchange are formed in the system. The isotherms of χ_{Mn} for the solid solutions with yttrium content 20 mol% are given in Fig. 3.12.

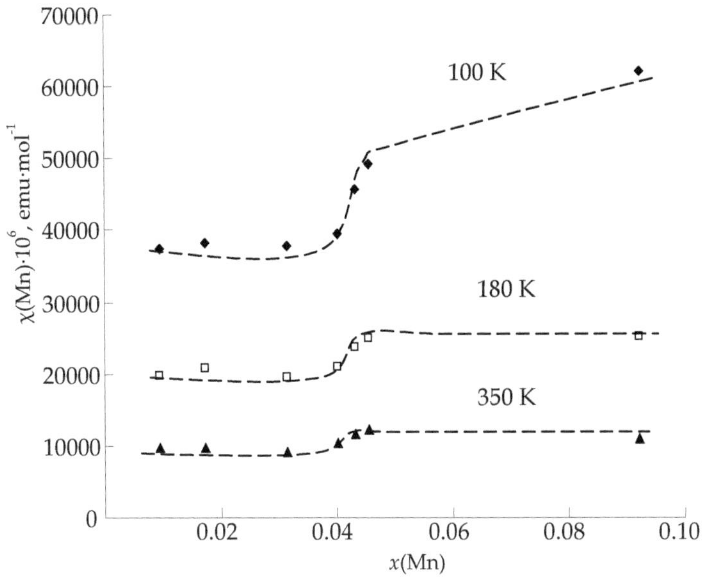

Figure 3.11 Plots of paramagnetic component of magnetic susceptibility vs. x for $x(La_{0.9}Y_{0.1})_{0.67}Sr_{0.33}MnO_3$–$(1-x)La_{0.9}Y_{0.1}AlO_3$ for three temperatures.

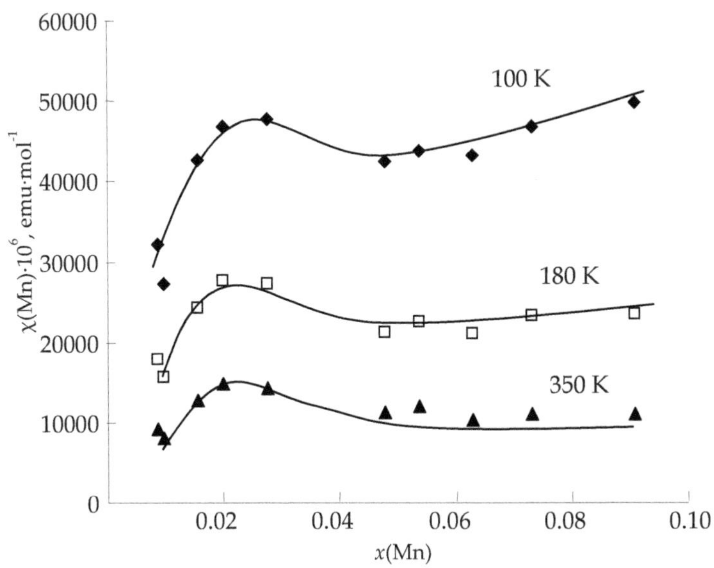

Figure 3.12 Plots of paramagnetic component of magnetic susceptibility vs. x for the $x(La_{0.8}Y_{0.2})_{0.67}Sr_{0.33}MnO_3$–$(1-x)La_{0.8}Y_{0.2}AlO_3$ for three temperatures.

In this case all the isotherms of paramagnetic component of magnetic susceptibility are characterized by a maximum and are similar in their shape to the yttrium–calcium series containing 20 mol% of yttrium. Such a behavior suggests that at x close to 0.02 a change in the character of magnetic exchange from ferro- to antiferromagnetic occurs in the system. Further increase in manganese atoms content higher than $x \sim 0.04$ results in a monotonous increase in magnetic susceptibility determining the prevalence of ferromagnetic clusters of manganese atoms.

A comparison between the isotherms of magnetic susceptibility for lanthanum manganites doped only with strontium $xLa_{0.67}Sr_{0.33}MnO_3$–$(1-x)LaAlO_3$ and with 10 and 20 mol% of yttrium results in the following. As is seen from Fig. 3.13 the χ_{Mn} isotherms are similar in their shape.

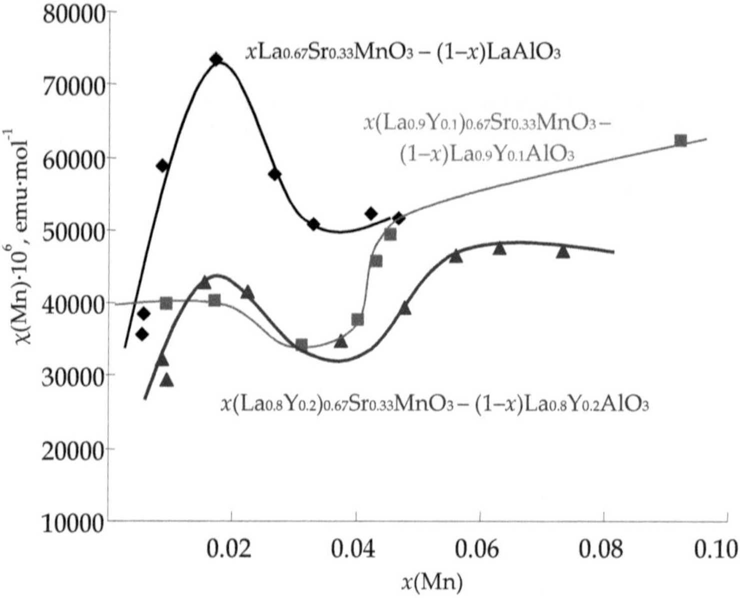

Figure 3.13 Isotherms of paramagnetic susceptibility at 100 K for various Sr-containing systems.

However, the changes in the magnetic properties of these manganites are not monotonous with the composition of diamagnetic sublattice. The solid solutions with 10 mol% of yttrium do not obey the general trend. The paramagnetic components of magnetic susceptibility in the systems containing 20 mol% of yttrium like in the case of $xLa_{0.67}Sr_{0.33}MnO_3$–(1–x)$LaAlO_3$ have a maximum at $x \sim 0.02$, which points to the existence of paramagnetic centers with prevailing antiferromagnetic exchange at $x > x_{max}$. Introduction of 10 mol% of yttrium into the lanthanum sites results in the absence of extremes in the χ_{Mn} isotherms and in the region of diluted solid solutions up to $x \sim 0.02$ the paramagnetic component of magnetic susceptibility is almost constant.

The effective magnetic moments at the infinite dilution in the case of the systems containing 10 mol% of yttrium decrease as the temperature increases from 5.98 μB at 77 K to 5.53 μB at 400 K. Such μ_{eff} ($x \to 0$) as in the case of yttrium–calcium manganites do not allow us to suggest the presence of single paramagnetic atoms at the infinite dilution. Among clusters of two manganese atoms the only possible situation is the formation of paramagnetic centers Mn^{3+}–O–Mn^{4+} of the composition $[Mn^{3+}Mn^{4+}YSrO_{11}]^{10-}$ and μ_{eff} ($x \to 0$) = 5.6 μB. In this case a moderate decrease in the effective magnetic moment at the infinite dilution as the temperature increases determines the ferromagnetic character of the exchange within clusters with a rather large exchange parameter $J > 90$ cm^{-1}.

Introduction of 20 mol% of yttrium into the perovskite structure results in μ_{eff} ($x \to 0$) ~ 3.55 μB almost independent of temperature. At first glance such a magnetic moment appears to be too low for any possible valence state of manganese atoms. Such low μ_{eff} ($x \to 0$) points to the presence of antiferromagnetic clusters in an infinitely diluted solid solution. An extremely approximate estimation of μ_{eff} ($x \to 0$) (owing to such a narrow concentration region) suggests the presence of single Mn^{3+} atoms and dimers Mn^{3+}–O–Mn^{3+} or Mn^{4+}–O–Mn^{4+} with strong antiferromagnetic exchange. As the concentration increases the clustering becomes more effective with constant competition of both types of exchange. In this case, when the extrapolation of magnetic susceptibility to the infinite dilution is hampered by an abrupt decrease in χ_{Mn}, we can consider the tendency to clustering by examination of experimental magnetic moments and their dependence on temperature for both strontium-containing systems.

Therefore, we found that the substitution of a fraction of lanthanum atoms for yttrium in $La_{0.67}Ca_{0.33}MnO_3$ and $La_{0.67}Sr_{0.33}MnO_3$ results in an increase in manganese atom clustering. For all the compositions $x(La_{1-z}Y_z)_{0.67}A_{0.33}MnO_3$–$(1-x)La_{1-z}Y_zAlO_3$ (A= Ca, Sr) even at the infinite dilution ($x \to 0$) no complete disaggregation of paramagnetic atoms occur. Moreover, the level of clustering is substantially greater in calcium containing systems than in the solid solutions with strontium.

The most important fact is that an increase in yttrium concentration affects the formation of exchange bonded aggregates of manganese atoms non-monotonously. The greatest aggregates were found to be formed upon introduction of 10 mol% of yttrium into the sublattice of heavy metals. Another important fact concerning the influence of yttrium nature on the state of paramagnetic atoms is that this tendency is observed both for the $x(La_{1-z}Y_z)_{0.67}Ca_{0.33}MnO_3$–$(1-x)La_{1-z}Y_zAlO_3$ and for $x(La_{1-z}Y_z)_{0.67}Sr_{0.33}MnO_3$–$(1-x)La_{1-z}Y_zAlO_3$ solid solutions.

The variation in the ratio of calcium and strontium atoms in the $xLa_{0.67}(Ca_ySr_{1-y})_{0.33}MnO_3$–$(1-x)LaAlO_3$ system also results in a non-monotonous changes in magnetic characteristics.

The use of only the size factors[28] to account for such a behavior of magnetic susceptibility is evidently inadequate. Antiferromagnetic character of exchange is known to result from orbital overlapping and must increase as the covalence of Mn–O bond increases. In contrast, ferromagnetic superexchange is determined by the interaction via two perpendicular *p*-orbital of oxygen atom and represents an electron correlation.[28] This means that ferromagnetic exchange must not be affected by the degree of ionicity of Mn–O bond.

Let us consider the effects of doping lanthanum manganite with Ca(Sr) and yttrium. The size of yttrium atoms are essentially less than the size of lanthanum, and doping with yttrium results in a decrease in the unit cell parameter. This means that the average Mn–O distance also decreases and the orbital overlapping increases, the absolute value of the parameter of antiferromagnetic exchange increases, and, as was shown in ref. 36, the clustering must also increase. On the other hand, a small yttrium atom polarizes *p*-orbitals of oxygen atoms to a greater extent resulting in an increase in the Mn–O bond ionicity and thus increasing the contribution of ferromagnetic exchange.

In other words the result of yttrium introduction may be represented schematically in Scheme 3.1.

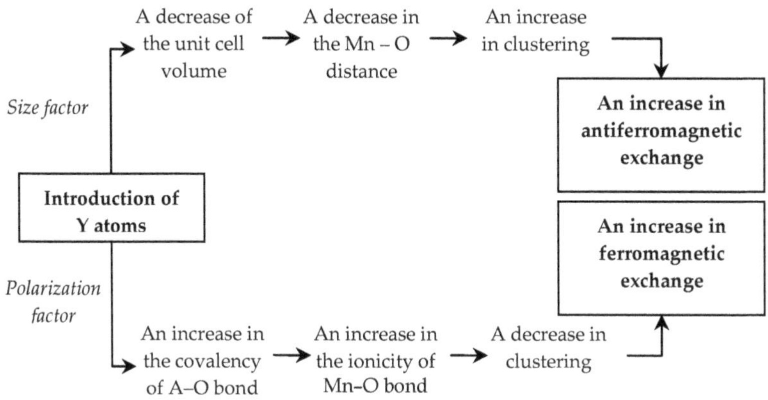

Scheme 3.1 Results of yttrium introduction in lanthanum aluminate doped with manganese.

Therefore, the introduction of yttrium into the lanthanum sites simultaneously results in opposing results. Moreover, at a certain yttrium concentration one of the tendencies appears to prevail. Given 10 mol% of yttrium, we have a maximum of clustering (size factor) and a maximum of ferromagnetic exchange (polarizing factor).

The difference in the behavior of the systems containing calcium and strontium seems to be associated with the fact that strontium being greater in size and forming a more ionic bond with oxygen in all the aspects given in the above presented scheme acts in opposite directions to yttrium. As the result, all the effects observed for calcium containing systems appear to be less pronounced, but show themselves in the same directions: a decrease in clustering as the concentration of yttrium increases and variations in the dominating characters of exchange as the concentration of paramagnetic atoms increases.

In the systems, where we vary the ratio Ca:Sr, both factors described above also are operative (Scheme 3.2).

As the ratio Sr:Ca increases, so does the size of the unit cell and Mn–O distance, the orbital overlapping decreases, and the ferromagnetic contribution to the exchange increases.

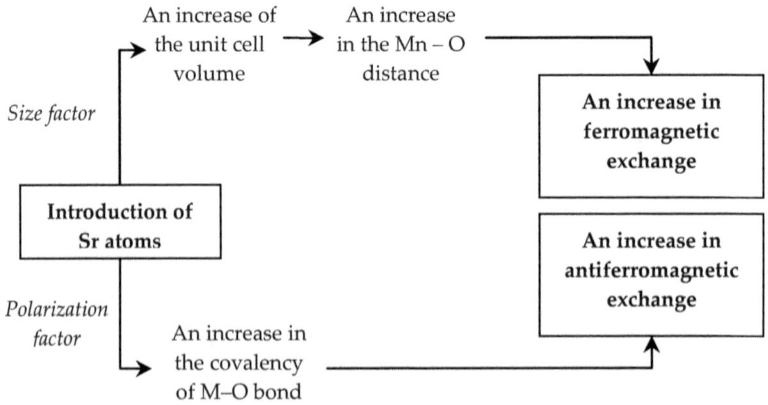

Scheme 3.2 Results of strontium introduction in lanthanum aluminate doped with manganese.

However, as the ionicity of A–O bond increases, the Mn–O bond becomes more covalent, which results in an increase in antiferromagnetic exchange. As the result, we found that both factors are balanced given the ratio 1:1. Moreover, an essentially greater role is played by the distortions of the nearest surrounding, and as a result, at low temperatures the effective magnetic moment attains greater values than for Sr:Ca = 0 or 1.

In other words, the main role in the effect of colossal negative magnetoresistance seems to belong to a certain balance, competition between ferromagnetic and antiferromagnetic interactions rather than to ferromagnetic exchange between manganese atoms with different oxidation states.

The contribution of both types of exchange to a large extent is determined by the nature of the doping elements—their sizes and polarizing ability. Importantly, the influence of doping elements shows up in the short order interactions—within small (4–8 atoms) clusters of manganese atoms located round or close to the diamagnetic doping elements.

Now two problems need to be ascertained: (1) How does the nature of rare earth element affect the clusterization of manganese atoms in the substituted lanthanum manganites? (2) What is the effect of partial substitution of yttrium for lanthanum and simultaneous variation of the ratio Sr:Ca.

To solve the first problem we synthesized and studied the solid solutions, where other rare earth elements replaced yttrium in the same ratios in the doped lanthanum manganites.

First of all, in the context of rare earth elements having unpaired electrons in the *f*-orbitals, we were to ascertain if they take part in the exchange interactions between themselves. Hence, the $La_{1-y}R_yAlO_3$ solid solutions (y = 0.01–0.20) (R = Ce, Eu, Yb, Gd) were synthesized and studied.[37-39] In all the cases the effect of magnetic dilution was observed—the paramagnetic component of magnetic susceptibility decreased as the rare earth concentration increased, which points to a weak but distinctive antiferromagnetic exchange between paramagnetic rare earth atoms. At the infinite dilution the effective magnetic moments could be well described by the single rare earth atom.

A comparison between magnetic properties of the solid solutions containing rare earth elements and containing yttrium in the same concentration in doped lanthanum manganites showed that the introduction of paramagnetic atoms of rare earth elements essentially affects the states of manganese atoms and the character of exchange interactions. This can be shown taking the $x(La_{0.9}Yb_{0.1})_{0.67}Ca_{0.33}MnO_3–(1–x)LaAlO_3$ and $xLa_{0.67}Ca_{0.33}MnO_3–(1–x)LaAlO_3$ as an example.[38] The dependences of paramagnetic components of magnetic susceptibility calculated per one mole of manganese atoms (with the components for ytterbium being subtracted) on manganese concentration substantially differ for these two systems (Fig. 3.14).

The effective magnetic moments at the infinite dilution of the $x(La_{0.9}Yb_{0.1})_{0.67}Ca_{0.33}MnO_3–(1–x)LaAlO_3$ solid solutions depend on temperature and vary from 5.67 μB (90 K) to 6.22 μB (320 K). The obtained μ_{eff} ($x \to 0$) corresponds to the formation of clusters containing three manganese atoms as a minimum. At the same time the absolute values of μ_{eff} ($x \to 0$) testify for ferromagnetic character of exchange, whereas the character of their temperature dependence is typical for the clusters with antiferromagnetic exchange. This points to a competition between ferromagnetic Mn^{4+}–O–Mn^{3+} exchange and antiferromagnetic exchange between manganese atoms in the same oxidation states. This means that the clusters must include at least three manganese atoms.

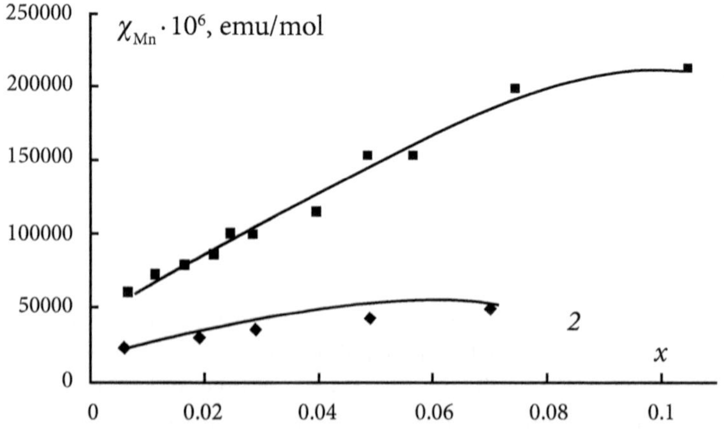

Figure 3.14 Plots of paramagnetic component of magnetic susceptibility calculated per 1 mole of manganese atoms vs. x for the $x(La_{0.9}Yb_{0.1})_{0.67}Ca_{0.33}MnO_3$–$(1-x)LaAlO_3$ (1) and $xLa_{0.67}Ca_{0.33}MnO_3$–$(1-x)LaAlO_3$ (2) solid solutions at $T = 100$ K.

The effective magnetic moments at the infinite dilution of the $x(La_{0.9}Yb_{0.1})_{0.67}Ca_{0.33}MnO_3$–$(1-x)LaAlO_3$ solid solutions depend on temperature and vary from 5.67 μB (90 K) to 6.22 μB (320 K). The obtained μ_{eff} $(x \to 0)$ corresponds to the formation of clusters containing three manganese atoms as a minimum. At the same time the absolute values of μ_{eff} $(x \to 0)$ testify for ferromagnetic character of exchange, whereas the character of their temperature dependence is typical for the clusters with antiferromagnetic exchange. This points to a competition between ferromagnetic Mn^{4+}–O–Mn^{3+} exchange and antiferromagnetic exchange between manganese atoms in the same oxidation states. This means that the clusters must include at least three manganese atoms.

Therefore, the effective magnetic moments and the isotherms of paramagnetic component of magnetic susceptibility for the $x(La_{0.9}Yb_{0.1})_{0.67}Ca_{0.33}MnO_3$–$(1-x)LaAlO_3$ solid solutions show an increase in the clustering of paramagnetic manganese atoms and also in the contribution of ferromagnetic type of exchange. The same is true for other rare earth elements introduced into lanthanum sites. The first reason for this is that the substitution of rare earth atoms having lower radius $r(Yb^{3+})_{XII}$ = 1.18 Å for lanthanum with $r(La^{3+})_{XII}$ = 1.36 Å[27] must result in local distortions of oxygen surrounding of manganese atoms and thus in some change in the Mn–O–Mn exchange angle. This favors an increasing ferromagnetic contribution into the exchange even between atoms in the same oxidation state. The second reason is that rare earth elements being smaller and having the same effective charge as lanthanum exert greater polarizing effect on oxygen orbitals, which results in an increase in the covalence of Mn–O bond and as was shown in ref. [28] in an increase in manganese clustering. However, for rare earth elements as doping atoms, the third and, perhaps, the most important reason (the other two are valid also for yttrium) consists of the following. All the data show that both rare earth elements and yttrium appear to be the components of the clusters of manganese atoms. However, they have unpaired electrons, which can take part even in the exchange between themselves and at a rather large distance, will inevitably contribute their electron density to the clusters, and such an exchange must necessarily be ferromagnetic.

As for the changes in the ratio of Ca:Sr given a fraction of yttrium in the lanthanum manganite, in an interesting work[40] the authors experimentally found that the greatest CMR effect is demonstrated by the $(La_{0.9}Y_{0.1})_{0.67}(Ca_{0.5}Sr_{0.5})_{0.33}MnO_3$ system.

We emphasize that the authors studied a wide series of compositions with various ratios between doping elements, but they tried to account for the observed effect only on the basis of comparing the sizes of doping atoms, i.e., leaning upon the size factor. Our studies described above show that this is not enough. Hence, with the aim to reveal any correlation between the nature and ratios of doping atoms we studied the magnetic dilution in the $(La_{0.9}Y_{0.1})_{0.67}(Ca_{0.5}Sr_{0.5})_{0.33}MnO_3$–$La_{0.9}Y_{0.1}AlO_3$ system.[38]

The isotherms of paramagnetic component of magnetic susceptibility (χ_{Mn}) for the systems with various diamagnetic compositions differ both by their shape and by the values of χ_{Mn} (Fig. 3.7).

As distinct from $xLa_{0.33}Ca_{0.67}MnO_3$–$(1-x)LaAlO_3$ solid solutions, introduction of yttrium into the perovskite structure results in a more complex shape of the χ_{Mn}–x isotherms and in the changes in the character of exchange interaction as the concentration decreases. It is interesting to note that the isotherms for the system in question—$x(La_{0.9}Y_{0.1})_{0.67}(Ca_{0.5}Sr_{0.5})_{0.33}MnO_3$–$(1-x)$

$La_{0.9}Y_{0.1}AlO_3$—have the same shape as the isotherms for $x(La_{0.8}Y_{0.2})_{0.67}Ca_{0.33}MnO_3$–$(1-x)La_{0.8}Y_{0.2}AlO_3$ (increased fraction of yttrium) and the isotherms for lanthanum manganite doped only with strontium (see Fig. 3.3). This is a good illustration of our schemes of the influence of yttrium on the one hand and of strontium on the other hand, although, of course, the values of χ_{Mn} differ for these systems. However, this shows that the most important changes in the sizes of clusters and in the exchange interactions in them become evident in the diluted solid solutions.

The effective magnetic moments in the $x(La_{0.9}Y_{0.1})_{0.67}(Ca_{0.5}Sr_{0.5})_{0.33}MnO_3$–$(1-x)La_{0.9}Y_{0.1}AlO_3$ solid solutions extrapolated to the infinite dilution decrease as the temperature increases from ~5.16 μB at 90 K to ~4.30 μB at 400 K. The obtained $\mu_{eff}(x \to 0)$ cannot be ascribed to single manganese atoms in any ratio. We again deal with some clusters preserved even at the infinite dilution. We carried out the calculation of magnetic susceptibility within the framework of Heisenberg–Dirac–Van Vleck model. On the basis of decreasing effective magnetic moment with temperature, we could assume with certainty that in the solid solutions we have the dimers of manganese atoms—Mn^{3+}–O–Mn^{4+}—and, perhaps, some fractions of single Mn^{3+} and Mn^{4+} atoms (see Chapter 1).

The calculation showed that the dependence of the effective magnetic moment at the infinite dilution for the system under study, with 10 mol% of yttrium and Ca:Sr = 1:1 is well described as a ferromagnetic dimer with $J = +7$ cm^{-1}. No single manganese atoms were found.[41]

A comparison between the effective magnetic moments at the infinite dilution for the systems with various diamagnetic compositions is given in Fig. 3.15.

For the systems containing only alkaline earth elements as the doping diamagnetic elements a complete disaggregation of manganese atoms is observed at the infinite dilution. Whereas for the systems containing also yttrium substantially large clusters of manganese atoms are found at $x \to 0$ ($n > 3$). Introduction of calcium and strontium atoms into the perovskite structure does not result in strong short order interactions between manganese atoms and in the formation of clusters stable till the infinite dilution. Introduction of yttrium increases the clustering of paramagnetic atoms to the extent that at $x \rightarrow 0$ rather stable dimer clusters Mn^{4+}–Mn^{3+} bonded ferromagnetically are preserved in the solid solutions. Varying the ratio Ca:Sr optimizes the short order interactions resulting in just dimer clusters with ferromagnetic exchange and sufficiently high energy of interatomic interactions. Magnetic interactions of predominantly ferromagnetic type resulting from an enlargement of these clusters seem to be responsible for the effect of colossal magnetoresistance in these systems. That is, it is safe to conclude that ferromagnetic dimers are the main structure unit determining the optimal characteristics of magnetoresistance in manganites containing yttrium, calcium, and strontium as doping elements. Their formation occurs in the systems containing 10 mol% of yttrium and the equimolar ratio Ca:Sr resulting in the composition $(La_{0.9}Y_{0.1})_{0.67}(Ca_{0.5}Sr_{0.5})_{0.33}MnO_3$.

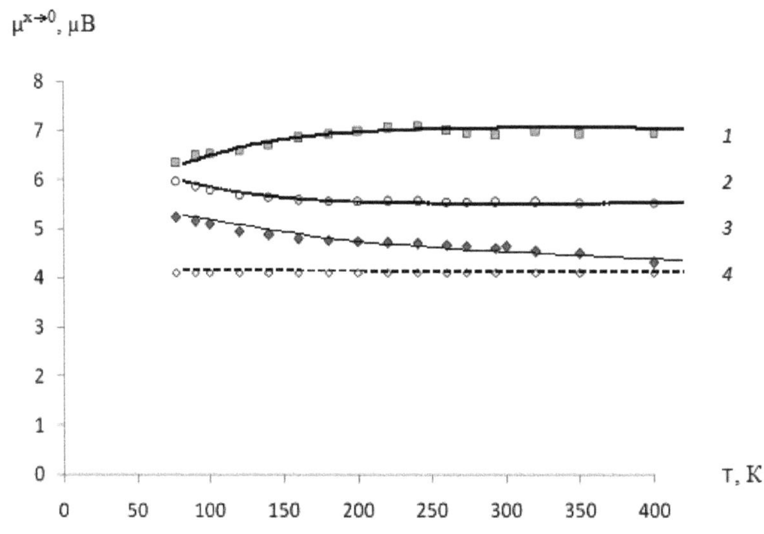

Figure 3.15 Plots of effective magnetic moments at the infinite dilution vs. temperature for the systems: (1) $x(La_{0.9}Y_{0.1})_{0.67}Ca_{0.33}MnO_3$–$(1-x)La_{0.9}Y_{0.1}AlO_3$; (2) $x(La_{0.9}Y_{0.1})_{0.67}Sr_{0.33}MnO_3$–$(1-x)La_{0.9}Y_{0.1}AlO_3$; (3) $x(La_{0.9}Y_{0.1})_{0.67}(Ca_{0.5}Sr_{0.5})_{0.33}MnO_3$–$(1-x)La_{0.9}Y_{0.1}AlO_3$; (4) $xLa_{0.33}Ca_{0.67}MnO_3$–$(1-x)LaAlO_3$.

It is quite evident that the fact that magnetic characteristics change no monotonously upon an increase in the Ca:Sr ratio with simultaneous introduction of yttrium furnish unambiguous proof of inefficiency of the size approach. In this case at least two factors may be operative, one of which is the degree of ionicity of Mn–O bond. The latter to a large extent is determined by the nature of the nearest neighbors in the lanthanum sublattice, i.e., the nature of the substituents.

Doping lanthanum manganite with a small yttrium atom results in a decrease in the unit cell parameter. In this case the average Mn–O distance decreases, the orbital overlapping increases, and, as a consequence, the absolute value of the parameter of antiferromagnetic exchange increases, so does the clustering of manganese atoms. On the other hand, a small yttrium atom polarizes the orbitals of oxygen atoms to a greater extent than lanthanum atoms thus resulting in an increase in the ionicity of M–O bond and in an increase in the contribution of ferromagnetic exchange. Introduction of strontium into the same lanthanum sites results in the opposite effect. In such a case at a certain concentration one of the tendencies appears to prevail.

For the composition studied here given 10 mol% of yttrium a maximal clustering must be observed with the greatest ferromagnetic exchange, but a simultaneous introduction of equal quantities of calcium and strontium results in essential changes in the short order interactions. This is another proof that first, such cooperative phenomena as colossal magnetoresistance are primarily determined by the short order interactions and, second, it is magnetic dilution that can reveal the character of these interactions.

References

1. Xu Y, Memmert U, Hartman U (2001) Magnetic field sensors from polycrystalline manganites, *Sens Actuators A*, **91**, 26–29.
2. Volger J (1950) Further experimental investigations on some ferromagnetic oxidic compounds of manganese with perovskite structure, *Physica*, **20**, 49–66.
3. Van Santen J, Jonker G (1950) Electrical conductivity of ferromagnetic compounds of manganese with perovskite structure, *Physica*, **16**, 599–600.
4. El-Fadli EMZ, Metni MR, Sapiña F, Folgado ABJV (2002) Electronic properties of mixed-valence manganates: The role of Mn substitutional defects, *Chem Mater*, **14**, 688–696.

5. Jonker G, Van Santen J (1950) Ferromagnetic compounds of manganese with perovskite structure, *Physica*, **16**, 337–349.
6. Jin S, Tiefel T, McCormack M (1994) Thousandfold change in resistivity in magnetoresistive La-Ca-Mn-O films, *Science*, **264**, 413–415.
7. Wollan EO, Koehler WC (1955) Neutron diffraction study of the magnetic properties of the series of perovskite-type compounds $La_{1-x}Ca_xMnO_3$, *Phys Rev*, **100**(2), 545–563.
8. Goodenough J (1955) Theory of the role of covalence in the perovskite-type manganites LaM(II)MnO_3, *Phys Rev*, **100**(2), 564–573.
9. Zener C (1951) Interaction between the *d*-shells in the transition metals. II. Ferromagnetic compounds of manganese with perovskite structure, *Phys Rev*, **82**(3), 403–405.
10. Pirogov AN, Teplykh AE, Voronin VI, Kar'kin AE, Balagurov AM, Pomyakushin VYu, Sikolenko VV, Petrov AN, Cherepanov VA, Filonova EA (1999) Ferro- and antiferromagnetic ordering in $LaMnO_{3+\delta}$, *Phys Solid State*, **41**(1), 91–96.
11. Hong CS, Kim WS, Chi EO, Lee KW, Hur NH (2000) Colossal magnetoresistance in $La_{0.7}Ca_{0.3}MnO_{3-\delta}$: Comparative study of single-crystal and polycrystalline material, *Chem Mater*, **12**(11), 3509–3515.
12. Trukhanov SV, Kasper NV, Troyanchuk IO (2002) Evolution of magnetic state in the $La_{1-x}Ca_xMnO_{3-\delta}$ manganites depending on the oxygen content, *J Solid State Chem*, **169**, 85–95.
13. Fesenko EG (1972) *Perovskite Family and Ferroelectricity*, Moscow, 245p. [In Russian].
14. Pickett W, Singh D (1996) Electronic structure and half-metallic transport in the $La_{1-x}Ca_xMnO_3$ system, *Phys Rev B*, **53**(3), 1146–1160.
15. Tomioka Y, Asamitsu A, Kuwahara H, Tokura Y (1999) Reentrant transition of the charge-ordered state in perovskite manganites, *J Phys Soc Jpn*, **66**(2), 302–305.
16. Tokura Y (1999) Colossal magnetoresistive manganites, *J Magn Magn Mater*, **200**, 1–23.
17. Millange F, Suard E, Caignaert V, Raveau B (1999) $YBaMn_2O_5$: Crystal and magnetic structure reinvestigation, *Mat Res Bull*, **341**, 1–9.
18. Zhu D, Zhu H, Zhang Y (2002) Hydrothermal synthesis of $La_{0.5}Ba_{0.5}$ MnO_3 nanowires, *Appl Phys Lett*, **80**, 1634–1636.

19. Barnabe A, Millange F, Maignan A (1998) Barium based manganites $Ln_{1-x}Ba_xMnO_3$ with Ln = Pr, La: Phase transition and magnetoresistance properties, *Chem Mater*, **10**, 252–259.

20. Radaelli P, Cox D, Marezio M (1997) Charge, orbital, and magnetic ordering in $La_{0.5}Ca_{0.5}MnO_3$, *Phys Rev B*, **55**(5), 286.

21. Wang YX, Du Y, Qin R (2001) Phase equilibrium of the La-Ca-Mn-O system, *J Solid State Chem*, **156**, 237–241.

22. Baikov YM, Nikulin EI, Meleh BT (2007) Conductivity of $Ce_xSr_{1-x}MnO_3$ manganites in a magnetic field within the temperature range 78–300 K, *Phys Solid State*, **49**(4), 730–733.

23. Damay F, Cohen LF, (2000) Low-temperature grain boundaries effect in $La_{0.7-x}Y_xCa_{0.3}MnO_3$, *J Magn Magn Mater*, **54**, 150–154.

24. Chezhina N, Mikhailova M, Osipova A (2001) Manganese reactivity in the synthesis of magnetoresisting complex oxides, *Solid State Ionics*, **141–142**, 617–621.

25. Chezhina NV, Fedorova AV, (2010) Influence of yttrium atoms on magnetic properties of lanthanum manganites doped with strontium, *Russ J Gen Chem*, **80**(2), 203–206.

26. Chezhina N, Kuzmich M (2004) Magnetic dilution in the $xLa_{0.33}Ba_{0.67}MnO_3$–$(1-x)LaAlO_3$ system, *Russ J Gen Chem*, **74**(4), 486–488.

27. Shannon RD, Prewitt CT (1969) Effective ionic radii, *Acta Cryst B*, **25**, 925–946.

28. Kalinnikov VT, Rakitin Yu V (eds) (1980) Introduction to magnetochemistry. Method of static magnetic susceptibility in *Chemistry*, Nauka, Moscow [In Russian].

29. Zhao S, Yue X-J, X, Liu X (2017) Influence of Sr doping on structural, electrical and magnetic properties of $La_{0.7}Ca_{0.3}MnO_3$ nanoparticles, *Ceram Int*, **43**(16), 13240–13246.

30. Zhao S, Yue X, Liu X (2017) Tuning room temperature T_p and MR of $La_{1-y}(Ca_{y-x}Sr_x)MnO_3$ polycrystalline ceramics by Sr doping, *Ceram Int*, **43**(5), 4594–4598.

31. Chezhina NV, Fedorova AV (2010) Short order in magnetoresistive lanthanum manganites doped with various diamagnetic elements, *Russ J Gen Chem*, **80**(5), 909–914.

32. Zhou W, Ma C, Cao M, Gan Zh, Wang X, Ma Y, Wang X, Tan W, Wang D, Du Y (2017) Large magnetocaloric and magnetoresistance effects in metamagnetic $Sm_{0.55}(Sr_{0.5}Ca_{0.5})_{0.45}MnO_3$ manganite, *Ceram Int*, **43**(10), 7870–7874.

33. Hamdi R, Tozri A, Smari M, Dhahri E, Bessais L (2017) Resistivity, I–V characteristics and Hall effect in $Dy_{0.5}(Sr_{1-x}Ca_x)_{0.5}MnO_3$ manganites, *Mater Res Bull*, **95**, 525–531.

34. Mansuri I, Varshney D, Kaurav N, Lu CL, Kuo YK (2011) Effects of A-site disorder on magnetic, electrical and thermal properties of $La_{0.5-x}Ln_xCa_{0.5-y}Sr_yMnO_3$ manganites, *J Magn Magn Mater*, **323**(3–4), 316–323.

35. Khelifi J, Dhahri E, Hlil EK (2014) The influence of disorder on the appearance of Griffiths phase and magnetoresistive properties in $(La_{1-x}Nd_x)_{2/3}(Ca_{1-y}Sr_y)_{1/3}MnO_3$ oxides, *Ceram Int*, **40**(1), 1641–1649.

36. Chezhina NV, Patii VP (2004) Interactions between Mn(IV) atoms in $Sr_2Mn_xTi_{1-x}O_4$ solid solutions, *Russ J Gen Chem*, **70**(9), 1425–1428.

37. Fedorova AV, Chezhina NV, Shilovskikh VV (2015) State of europium atoms and exchange interactions in $La_{1-y}Eu_yAlO_3$, *Russ J Gen Chem*, **85**(10), 2223–2226.

38. Fedorova AV, Chezhina NV, Sukhenko KYu (2016) Magnetic properties of solid solutions of lanthanum manganite doped with ytterbium and calcium in $LaAlO_3$, *Russ J Gen Chem*, **86**(7), 1552–1557.

39. Ponomareva EA, Fedorova AV, Chezhina NV (2017) Magnetic susceptibility of $La_{1-y}Ce_yAlO_3$ solid solutions, *Russ J Gen Chem*, **87**(11), 2730–2732.

40. Cai HL, Wu XS, Wang FZ, Hu A, Jiang SS, Gao J, Tan WS (2005) A-site disorder induces magnetoresistance in Y and Sr co-doped $La_{2/3-x}Y_xCa_{1/3-y}Sr_yMnO_3$, *J Alloys Compd*, **397**(1–2), 250–254.

41. Chezhina NV, Fedorova AV (2014) Interatomic interactions in lanthanum manganite doped with yttrium, calcium, and strontium ($La_{0.9}Y_{0.1}Ca_{0.5}Sr_{0.5}MnO_3$), *Russ J Gen Chem*, **84**(12), 2382–2387.

Chapter 4

Influence of the Nature and Concentration of Dia- and Paramagnetic Elements on Electron Structure and Electrophysical Properties of Doped Lanthanum Gallate

Dmitry A. Korolev and Natalia V. Chezhina

Department of General and Inorganic Chemistry, St. Petersburg State University, Universitetskaya nab. 7/9, St. Petersburg, 199034, Russia

chemdim@mail.ru, d.korolev@spbu.ru, n.chezhina@spbu.ru

4.1 Introduction

Lanthanum gallate doped with strontium or simultaneously with strontium and magnesium, and also with transition elements those latter entering the sites of gallium in perovskite structure, opened a wide series of oxygen ionic and electron-ionic conductors promising for energy-saving technologies in the so-called solid oxide fuel cells (SOFC).

Simultaneous doping of lanthanum gallate with alkaline-earth elements and magnesium makes it possible to obtain a wide

Electronic Structure of Materials: Challenges and Developments
Edited by Natalia V. Chezhina and Dmitry A. Korolev

ISBN 978-981-4800-55-6 (Hardcover), 978-0-429-24287-8 (eBook)
www.panstanford.com

spectrum of compounds with ionic conductivity. Introduction of transition elements into the gallium sites, on the whole, contributes to the appearance of electronic component of conductivity. An active search for the compositions providing for maximal ionic and (or) electronic conductivity is going on since 1994, when for the first time the ionic conductivity was found in lanthanum gallate doped with strontium, calcium, barium, and magnesium.[1]

This is accounted for by practical importance of the materials based on doped lanthanum gallates in the production of solid oxide fuel cells (SOFC).

The tolerance of perovskite structure of lanthanum gallate to various substitutions allows a wide spectrum of compositions to be obtained, which have various qualitative and quantitative composition and, consequently, various properties. Doped gallates are used as a rule as electrolytes in SOFC.[2–7]

In such a case the cathodes must have both electronic and ionic conductivity and the electrolytes—purely ionic. $LaGaO_3$ doped with strontium and magnesium, as an electrolyte, is now a popular subject for SOFC operating in the middle temperature region (600–800°C).

A large number of works are devoted to the study of the structure of lanthanum gallate and its derivatives.[8–22]

The data on structural analysis are ambiguous, which is substantially determined not only by the synthetic procedure but also by the quality of starting reagents.

The study in ref. 9 is particularly remarkable, since the structure of $La_{1-x}Sr_xGa_{1-y}Mg_yO_{3-\delta}$ is determined as a function of inserted Sr and Mg: orthorhombic at $x + y < 0.25$; orthorhombic and rhombohedral at $0.25 < x + y < 0.30$; at $x + y > 0.30$, if x or $y \geq 20$—cubic. By various data the structural phase transitions in pure and doped lanthanum gallate can occur over a wide range of temperatures.[16,18,20–22]

The choice of strontium and magnesium as heterovalent dopants is not casual, though the greater stability of Sr- and Mg-doped gallate compared to Ba- and Ca-containing analogs was found experimentally.[1] The thermodynamic examination shows a lesser endothermic effect of the solid solution formation on the introduction of Sr (138 ± 12 kJ·mol^{-1} for $La_{1-x}Sr_xGa_{1-y}Mg_yO_{3-\delta}$ and 166 ± 12 kJ·mol^{-1} for $La_{1-x}Ba_xGa_{1-y}Mg_yO_{3-\delta}$; $x > 0$, $y \leq 0.2$) or magnesium (275 ± 37 kJ·mol^{-1} for $LaGa_{1-y}Mg_yO_{3\delta}$; $y \leq 0.2$).[23]

This suggests that Sr and Mg are the most suitable doping elements in lanthanum gallate from the elements of the II group of the Periodic Table.

A number of works on computer simulation are devoted to the stability of doped lanthanum gallate and its structure.[24-26] The data of Khan using the statistic lattice simulation method[24] are in good agreement with endothermic character of experimental values in ref. 23.

It is interesting to note that by the data of computer simulation the formation of oxygen vacancies upon heterovalent substitution of lanthanum gallate results in its destabilization, whereas doping zirconium oxide with yttrium contributes to the stabilization of the fluorite structure.[23] In Sr- and Mg-doped gallate the binding energy of "bivalent cation–vacancy" for a $Mg_{Ga}'|V_O$ couple was found to be –0.90 kJ·mol^{-1}, for a $Sr_{La}'|V_O$ couple—approximately –0.01 kJ·mol^{-1},[25] which can be a promising factor favoring an increase in the oxygen conductivity on introduction of strontium. The data of calculations[24] are in agreement with experimental transport numbers (t_O) in gallates doped with strontium and magnesium, where a decrease in the oxygen transport numbers is noted for the systems containing only strontium compared to magnesium-containing solutions (t_O = 0.98 for $La_{0.9}Sr_{0.1}GaO_{3-\delta}$ and 0.92 for $LaGa_{0.85}Mg_{0.15}O_{3-\delta}$ at 1073 K).[27]

Attention must be paid to a serious problem upon doping lanthanum gallate with alkaline-earth metals and magnesium only—the obtained samples are not single phase. Again in this case much depends on the conditions and methods of the synthesis and also on the very precursors.

Aside from the phase of gallate, many researchers found admixing phases such as $LaSrGaO_4$ and $LaSrGa_3O_7$,[8-11,27-38] in all cases where the solubility of strontium is exceeded or close to the limit. By the data of ref. 1, the limit of strontium solubility in $LaGaO_3$ is 10 at%. It is interesting to note that introduction of magnesium only up to 20 at% does not result in admixture phases.[28] An excess of magnesium results in the isolation of $La_4Ga_2O_9$, $LaSrGaO_4$, and $LaSrGa_3O_7$. As a rule the quantity of admixture phases is small and attains no more than 5%. The reason for their appearance is the instability of the structure of doped gallate, which, as has been shown above, can be accounted for by thermodynamics. That means that in general from energetic point of view the emergence of oxygen vacancies is unfavorable. The admixture phases being insulating to a large extent results in a decrease in the conductivity and in a decrease in the life time of a material. In some cases the admixture compounds were not detected.[39]

At the same time, in some works the effects of the stabilization of the structure of lanthanum gallate doped with Sr and (or) Mg were found upon the introduction of transition element ions.[40,41] In the studies of conductivity much attention is being given to lanthanum gallates doped apart from strontium and magnesium with transition elements.[4,27,34,35,42–54] In this case we deal with electron-ionic conductors, which are interesting not only from applied point of view, but also from the point of view of fundamental studies. The main intriguing problem in this case is the separation of electronic and ionic components of conductivity, which is sometimes a very complicated problem. For elucidating the qualitative and quantitative regularities in the systems in question, the most important seems to be the knowledge of their electron structure. This includes not only the valence and spin states of the atoms, but also their mutual influence and the special features of interatomic interactions. The changes in the type of conductivity on varying the qualitative or quantitative composition of conductors is directly associated with interatomic interactions and atom states in their structure. Up to now the question about the electron structure of doped lanthanum gallate remains open, being partially solved on using the quantum chemistry calculations. There is an opinion that the introduction of strontium together with a transition metal into $LaGaO_3$ results in a partial transition of trivalent element into a higher valence state, usually +4[27,43,44,55] and sometimes even +5.[45]

The conclusion about the change in the valence state of a transition element is made from indirect data: a decrease in the volume of the unit cell (since cations with greater charge have a smaller ionic radius[56]),[44] or an increase in the activation energy of ionic conductivity (compared to $La(Sr)Ga(Mg)O_{3-\delta}$), which is accounted for by strong Coulomb interactions between tetravalent metal cations and oxygen ions in the M^{4+}–O–Mg^{2+} or M^{2+}–O–M^{4+} clusters.[27] However, such explanations are open to question since the changes in the structural parameters may be associated also with clustering and with the formation of oxygen vacancies. The conclusion about the emergence of cations with the +4 charge on introduction of Sr and Mg is questionable also because in this case the conductivity has to decrease or totally disappear; however, no such phenomena are observed. Moreover, the disproportionation of trivalent cations is very unprofitable (except for Mn^{3+}) for thermodynamic reasons. There is no unambiguous data directly pointing to an increase in the oxidation state of transition element atoms.

These rather urgent problems require unambiguous answers. In spite of a large body of data on this topic in their great majority the studies are empirical and random. The absence of systematic approach to the study of electron-ionic conductors within the context of "composition–structure–property" hampers the interpretation of experimental results and does not permit their generalization. The data about the electron structure would allow an insight into the problem of electronic and ionic conductivity of doped lanthanum gallates.

Several problems arise in selecting the qualitative and quantitative compositions of such electron-ionic conductors:

1. What is the difference between lanthanum gallate and aluminate as a matrix for electron ionic conductors? Would aluminate be much less expensive?
2. What must be the ratio between dia- (Sr and/or Mg) and paramagnetic (transition metal atoms) dopants to obtain the best electrophysical characteristics?
3. Which transition element would give the maximal ionic conductivity (a material for electrolytes) and which would result in the mixed electron-ionic conductivity (a material for cathodes)?
4. What is the reason for the stabilizing effect of a transition element, if its transfer to the higher oxidation state would eliminate vacancies in the oxygen sublattice and hamper the ionic conductivity?
5. Why was strontium empirically selected as a doping element in the lanthanum sites instead of calcium or barium?

The solution to all these problems lies in the electron structure of doped lanthanum gallates.

We emphasize that electron-ionic conductors based on doped lanthanum gallate are just rather diluted by paramagnetic elements solid solutions. Therefore, the systematic study of magnetic dilution must be the method able to reveal their electron structure and answer the questions formulated above. With this aim, a detailed study of several systems was carried out. First we undertook the study of the $La_{1-0.2x}Sr_{0.2x}M_xGa_{1-x}O_{3-\delta}$ solid solutions with various transition metals and fixed ratio M:Sr = 5:1. The data obtained were compared to the systems containing no strontium and with the solid solutions based on $LaAlO_3$. The next step consisted in changing the ratio M:Sr = 2:1 and in introducing magnesium into gallium sites—$La_{1-0.5x}Sr_{0.5x}M_xGa_{1-x}O_{3-\delta}$, $LaM_xGa_{1-1.2x}Mg_{0.2x}O_{3-\delta}$, $LaM_xGa_{1-1.5x}Mg_{0.5x}O_{3-\delta}$, $La_{1-0.2x}Sr_{0.2x}M_xGa_{1-1.2x}Mg_{0.2x}O_{3-\delta}$. We also studied the solid solutions containing calcium and barium as a doping element. In a number of cases we studied the electrical properties of obtained solid solutions to correlate the data of the electron structure with electrical performance.

4.2 Synthesis and Material Characterization

4.2.1 Synthesis

All the solid solutions under study were obtained by ceramic procedure. In some cases, for example to obtain the complex solid solutions of the type of $La_{1-0.2x}Sr_{0.2x}M_xGa_{1-1.2x}Mg_{0.2x}O_{3-\delta}$ the sol-gel method was used. The starting substances were: Special pure grade La_2O_3, Ga_2O_3, Mn_2O_3, and MgO (99.999%), analytical pure grade $SrCO_3$ (99.99%), NiO, and CoO (99.995%) were obtained from analytical pure grade metal nitrates, chromium oxide—by thermal decomposition of $(NH_4)_2Cr_2O_7$. All the starting substances were checked for the absence of ferromagnetic impurities by magnetic susceptibility method.

The ceramic procedure included a thorough grinding of the mixture of starting oxides and carbonates in an agate mortar, pelleting, and sintering at 1450°C. The time of sintering (50 h) was determined by the data of X-ray analysis and from the measurements of magnetic susceptibility after 40, 50, and 60 h. The magnetic susceptibility remained constant after 50 h of sintering, which pointed to the fact that the distribution of dia- and paramagnetic doping elements in the lattice of $LaGaO_3$ was close to the equilibrium.

The cobalt- and nickel-containing solid solutions were in addition sintered in flowing oxygen for 10 h, and the susceptibility being unchanged proved that the state of Co and Ni was constant.

The sol-gel method consisted of dissolving the oxide mixture in diluted HNO_3, adding citric acid and ethylene glycol to the solution neutralized to pH ~ 7 with ammonium hydroxide. The highly dispersed powder was obtained after a slow (4°/min) decomposition of the obtained gel, it was pressed into pellets and sintered in air for 50 h at 1450°C. With the aim of being able to compare the data of both methods in every oxide system, several samples were obtained by sol-gel method in addition to the ceramic procedure. The magnetic results coincided for both synthetic methods.

4.2.2 Methods of Material Characterization

The X-ray analysis was performed for every sample under study. The X-ray patterns were recorded on a Bruker D2 Phaser diffractometer, CuK_α emission.

In the case of nickel-containing solid solutions the anomalous thermal behavior of effective magnetic moments made us to study the temperature dependence of the structure of doped lanthanum gallate.

Since the content of paramagnetic element can change during the sintering the chemical analysis of paramagnetic and diamagnetic doping elements was performed for all the solid solutions by method of atom emission spectroscopy with the inductive bonded plasma. The error of the determination of x in the solid solution formula did not exceed 3%.

We measured the specific magnetic susceptibility of the solid solutions by Faraday method in the temperature range 77–400 K. The accuracy of relative measurements of specific magnetic susceptibility was 1%. We calculated the paramagnetic component of magnetic susceptibility. The diamagnetic corrections were introduced with regard to the susceptibility of lanthanum gallate matrix measured in the same temperature range as the samples under study.

The electrical properties of the solid solutions were studied by the method of impedance spectroscopy on an Impedancemeter-Z3000 in the temperature range 25–800°C and the frequencies from 1 MHz to 100 Hz.

We analyzed the impedance hodographs, separated the contributions of the volume and grain boundary components of the conductivity. The hodographs were interpreted in terms of the block approach with two equivalent electric schemes for low and high temperature regions. In all the cases we observed lower values of grain boundary component of conductivity at low temperatures and their leveling in the high temperature range. The obtained values were plotted as the dependencies of $lg\sigma - 1/T$.

4.3 Lanthanum Gallate Doped with Transition Element and Strontium

4.3.1 Lanthanum Gallate Doped with Transition Element and Strontium in Ratio M:Sr = 5:1

Since on doping lanthanum gallate with strontium and transition elements the ratio M:Sr = 5:1 is the most extensively used, we studied the solid solutions $La_{1-0.2x}Sr_{0.2x}M_xGa_{1-x}O_{3-\delta}$, varying 3*d*-element—Cr, Mn, Co, Ni. The data on magnetic dilution were compared with the results for the solid solutions containing no strontium—$LaM_xGa_{1-x}O_3$.

The plots for each 3*d*-element in question show the isotherms of the paramagnetic part of the magnetic susceptibility χ_M–x for the solid solutions with and without strontium additives (Fig. 4.1).

4.3.1.1 Chromium-containing systems

First we compared the data on lanthanum gallate with lanthanum aluminate and used the chromium-containing systems.

The reason was that Cr^{3+} is in $^4A_{2g}$ ground state, which makes the calculations within HDVV model the most unambiguous. The isotherms of paramagnetic component for both systems appeared to be similar (Fig. 4.2), but the plots of χ_{Cr} for gallate lie higher than for aluminate. The calculation of the fraction of dimer clusters showed (Fig. 4.3) that clustering in gallates is greater with a lower exchange parameter J (-12 cm^{-1} in $LaGaO_3$ vs. -18 cm^{-1} in $LaAlO_3$).[70] The calculation of the mixing energy in both matrices based on the fraction of dimer clusters[61,62] by the following equation:

$$\frac{\left(x-\frac{a_2x}{2}\right)}{\frac{a_2x}{z}\left(1-2x+\frac{a_2x}{z}\right)}=\exp\left(-\frac{2W_{12}}{kT}\right),$$

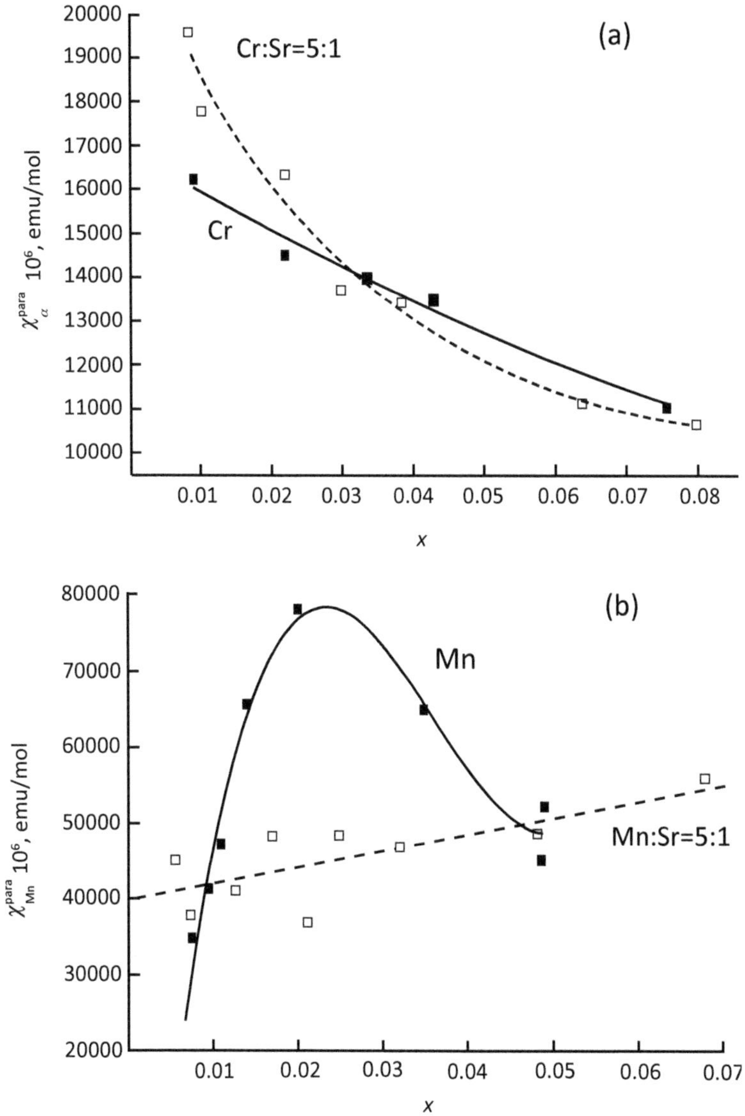

Figure 4.1 (*Continued*).

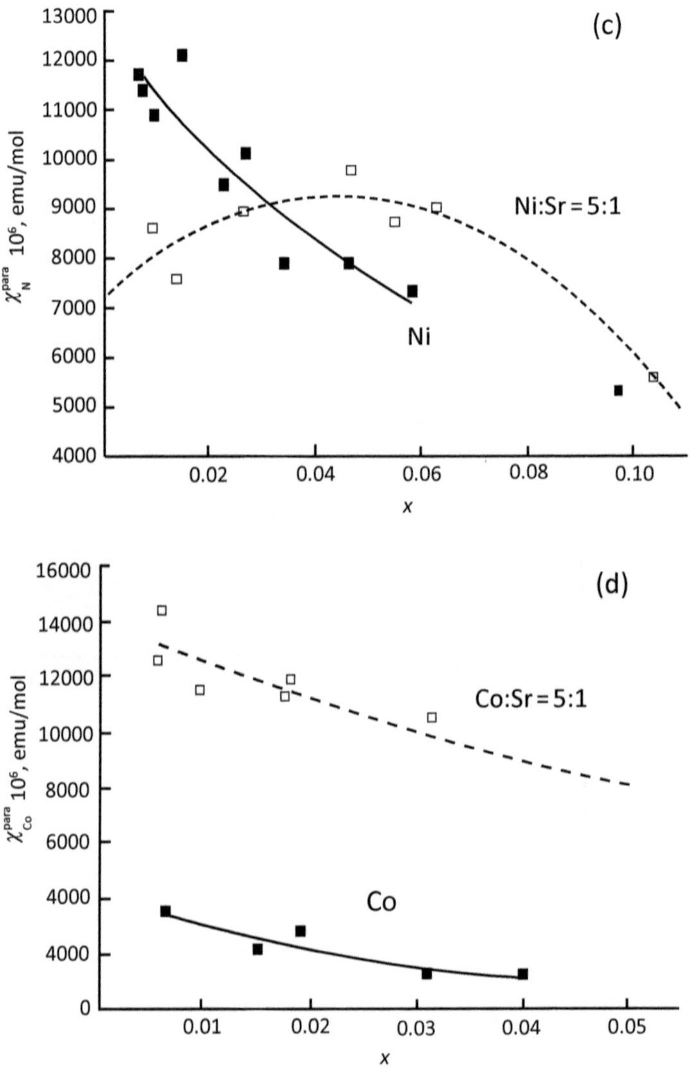

Figure 4.1 (a, b, c, d) The dependencies of paramagnetic components of magnetic susceptibility on transition metal content for the $LaM_xGa_{1-x}O_3$ and $La_{1-0.2x}Sr_{0.2x}M_xGa_{1-x}O_{3-\delta}$ for 140 K.

where x is the concentration of dopant, a_2 the fraction of dimers, W_{12} the mixing energy of solid solution formation, z the coordination number for dopant (z = 6 for perovskite), and W_{12} 12 $kJ \cdot mol^{-1}$ for aluminate and 15 $kJ \cdot mol^{-1}$ for gallate solid solution.

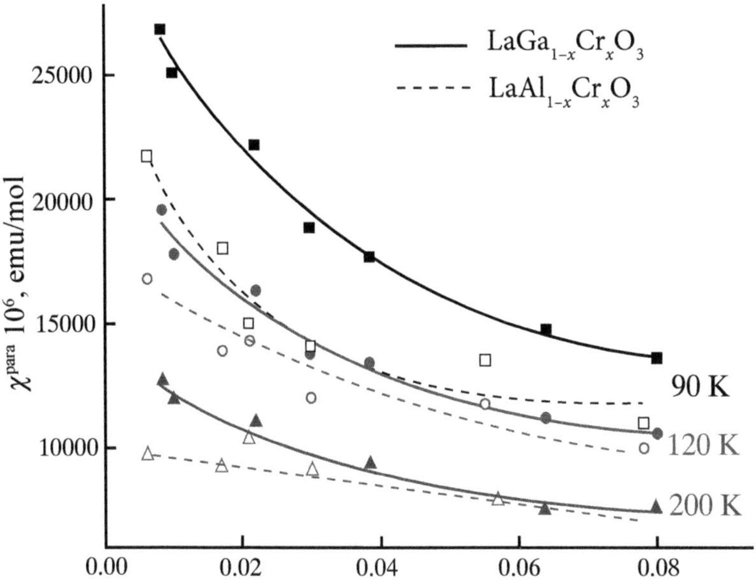

Figure 4.2 Dependencies of paramagnetic component of magnetic susceptibility on chromium concentration for Cr-containing aluminate and gallate systems.

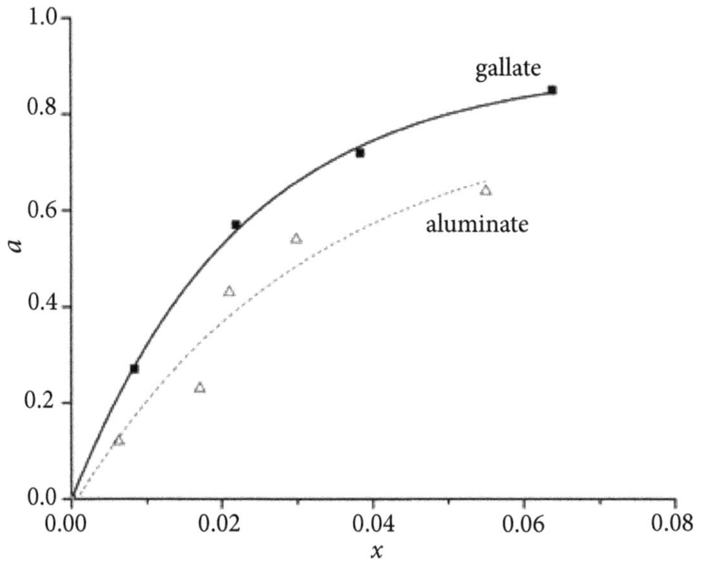

Figure 4.3 Dependencies of the fractions of dimer clusters in lanthanum aluminate and lanthanum gallate systems.

Therefore, we found that on passing from aluminate to gallate the clustering increases even without any diamagnetic heterovalent dopants. It is obvious that the enhancement of the aggregation in gallates is attributed to the increase in the covalent character of the M–O bond due to the reduced polarizing ability of gallium as compared to aluminum. The weakening of antiferromagnetic exchange is caused by the orthorhombic distortions in the gallate structure, which decrease the orbital overlapping.

Except for the chromium-containing samples, the isotherm patterns and the absolute values of the magnetic susceptibility significantly differ for the solid solutions $LaM_xGa_{1-x}O_3$ (M) and $La_{1-0.2x}Sr_{0.2x}M_xGa_{1-x}O_{3-\delta}$ (M:Sr = 5:1). Taking into account that the most concentrated solution contains only two Sr atoms per 98 La atoms, this circumstance is a crucial factor in the speculations on the causes of the structure stabilization.

First of all, let us look on the magnetic data for the solid solutions containing chromium. The fit of the magnetic susceptibility isotherms for the solid solutions $LaCr_xGa_{1-x}O_3$ and $La_{1-0.2x}Sr_{0.2x}M_xGa_{1-x}O_{3-\delta}$ within the entire concentration interval at $x > 0.02$ (Fig. 4.1a) indicates that the chromium atoms are in the same state in both systems. Partial oxidation of Cr^{3+} to Cr^{4+} should result in a decrease in magnetic susceptibility (Cr^{3+}–d^3, Cr^{4+}–d^2), as it has been observed earlier for perovskite-like structures.[63]

However, at very low concentrations of chromium the isotherm for the solid solutions containing strontium sharply goes up (over the whole temperature range). The effective magnetic moment for the infinite dilution ($x \to 0$) is higher than the spin only value for Cr^{3+} atom and increases with temperature.

This circumstance does not allow us to attribute it to the reduction of chromium to Cr^{2+}–^5E_g ground term. This may suggest that at infinite dilution in the systems containing strontium, disaggregation of chromium atoms is incomplete. The presence of clusters of chromium atoms at $x = 0$ indicates, first of all, that the energy of interatomic interactions in them is fairly high, more than 100 $kJ \cdot mol^{-1}$.[64] Emergence of such stable clusters in perovskite upon their doping with strontium along with a transition element indicates that strontium atoms are included into the clusters as, consequently, are the vacancies in the oxygen sublattice bonded with the strontium atoms. It is obvious that the appearance of clusters including transition element and strontium atoms and oxygen vacancies is responsible for the stabilization of the perovskite structure upon heterovalent doping. At this stage it appeared impossible to calculate the susceptibility of chromium-containing system within the model of diluted solution. This will be done later.

4.3.1.2 Manganese-containing systems

For the manganese-containing systems, the extrapolation of χ_{Mn} (Fig. 4.1b) and μ_{eff} to the infinite dilution ($x \to 0$) of the solid solutions $LaMn_xGa_{1-x}O_3$ and $La_{1-0.2x}Sr_{0.2x}Mn_xGa_{1-x}O_{3-\delta}$ gives the following values: $\mu_{eff} \sim 4.90$ μB (Mn^{3+}) and $\mu_{eff} \sim 6.14$ μB, respectively.[65]

For $x = 0$, χ_{eff} for the system containing Sr corresponds to the presence of only ~ 14% of Mn^{3+} monomers and ~ 86% of dimers, obviously, Mn^{2+}–O –Mn^{4+} ($\mu_{eff} = 6.32$ μB, given $J > 100$ cm^{-1}).

The value of $\mu_{eff} \sim 8.49$ μB in the region of the maximum in the isotherm for $LaMn_xGa_{1-x}O_3$ is close to the value of μ_{eff} for the ferromagnetic tetramer Mn^{2+}–O–Mn^{4+}–O–Mn^{2+}–O–Mn^{4+} (the ferromagnetic exchange can occur during disproportionation of Mn^{3+}).

Further decrease in the magnetic moment with the decreasing concentration indicates the appearance of antiferromagnetic interactions between tetramers along the axis z of $LaGaO_3$ unit cell.

The calculation of magnetic characteristics for the $LaMn_x Ga_{1-x}O_3$ system (Fig. 4.4a) showed that the amount of tetramers with ferromagnetic exchange ($J \geq 40$ cm^{-1}) increases up to $x \sim 0.02$, with a sharp decrease in the content of dimers with $J \geq 100$ cm^{-1}. As the concentration increases, tetramers with antiferromagnetic exchange ($J \leq -30$ cm^{-1}) begin to make significant contribution.

In the case of strontium-containing solid solutions, the calculation showed that the amount of monomers and dimers ($J \geq 100$ cm^{-1}) smoothly decreases with the increase in concentration, whereas the amount of tetramers increases ($J \geq 40$ cm^{-1}) (Fig. 4.4b).

It is obvious that the introduction of strontium into manganese-containing lanthanum gallate results in the formation of very strong clusters formed from the manganese atoms and including strontium atoms and vacancies, which do not disintegrate even upon infinite dilution.

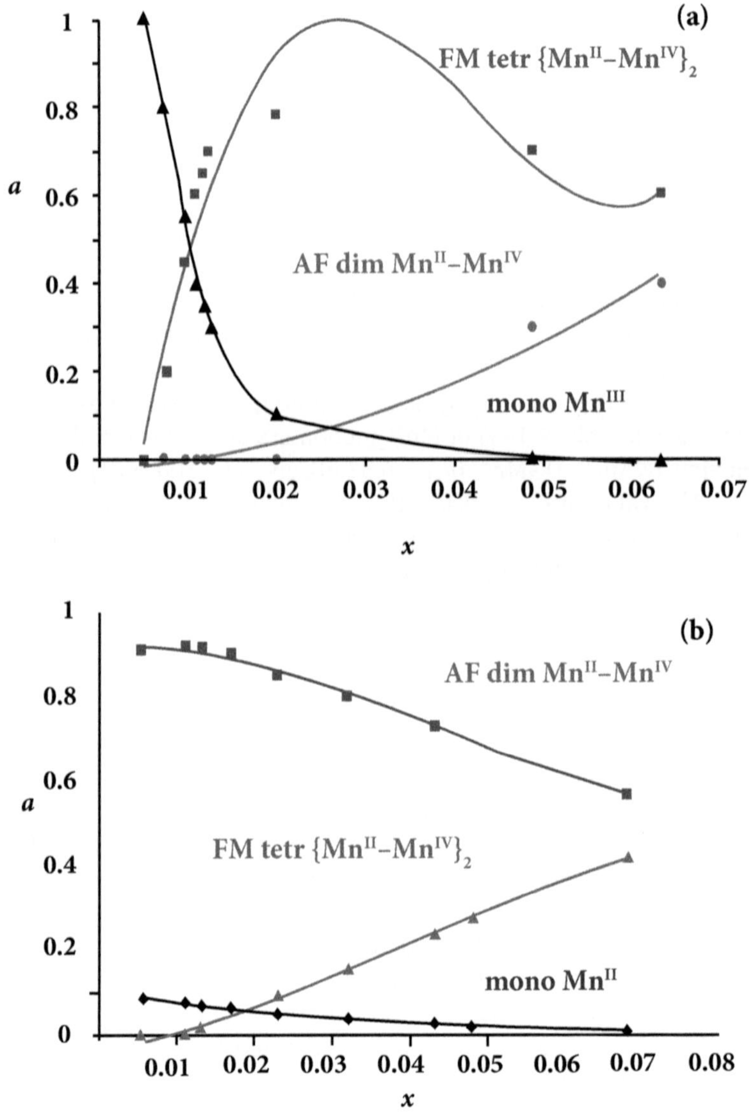

Figure 4.4 (a, b) Dependence of the aggregate contents on the concentration of manganese in solid solutions $LaMn_xGa_{1-x}O_3$ (a) and $La_{1-0.2x}Sr_{0.2x}Mn_xGa_{1-x}O_{3-\delta}$ (b).

4.3.1.3 Cobalt-containing systems

For the systems containing cobalt and nickel, the situation is more complicated because of a possibility of realization for these elements of high and low-spin states. The nonzero values of χ_{Co} for Co^{3+} and the values of effective magnetic moment at infinite dilution for nickel suggest the existence of spin equilibrium (see Chapter 1).

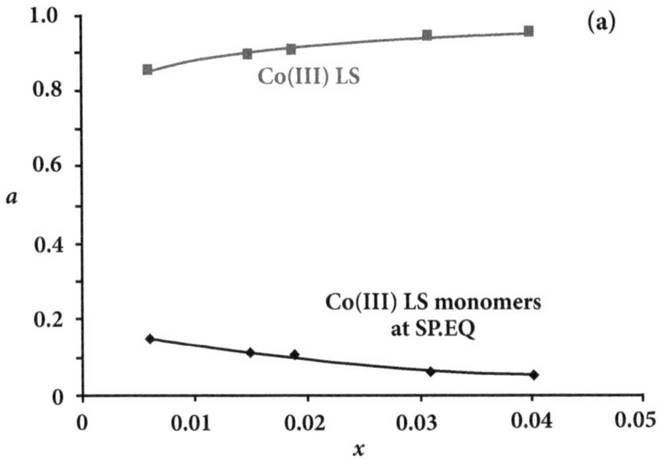

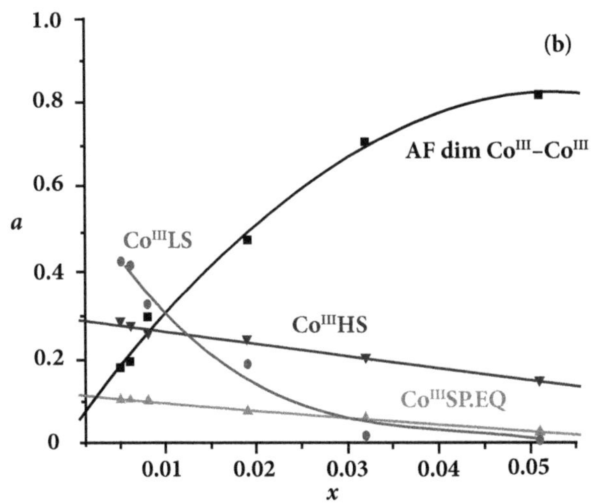

Figure 4.5 (a, b) Dependence of the aggregate contents on the concentration of cobalt in solid solutions $LaCo_xGa_{1-x}O_3$ (a) and $La_{1-0.2x}Sr_{0.2x}Co_xGa_{1-x}O_{3-\delta}$ (b).

When χ_{Co} (Fig. 4.1d) and μ_{eff} are extrapolated to infinite dilution, in the case of $LaCo_xGa_{1-x}O_3$ μ_{eff} increases from 1.90 to 2.43 μB, the values of μ_{eff} for $La_{1-0.2x}Sr_{0.2x}Co_xGa_{1-x}O_{3-\delta}$ are 3.32–3.87 μB.[68] The calculation showed that for $x = 0$ in the gallate doped with cobalt alone, only 20% of cobalt are in the spin state equilibrium $^1A_{1g} \leftrightarrow {}^5T_{2g}$, whereas 80% are in the low-spin state.

For $La_{1-0.2x}Sr_{0.2x}Co_xGa_{1-x}O_{3-\delta}$ a similar calculation showed that at infinite dilution ~14% of cobalt is in the spin equilibrium, ~56% in the low-spin state, and ~30% in the high-spin state.

The calculation of magnetic characteristics for $LaCo_xGa_{1-x}O_3$ showed the absence of high-spin dimers (Fig. 4.5a). Upon doping with strontium, as the concentration increases cobalt transforms to the high-spin state and forms dimers with the exchange parameter $J \sim -80$ cm^{-1} (Fig. 4.5b).

In the ESR spectra of the solid solutions under study, only one line with $g = 4.3$ was observed, which corresponded to high-spin Co^{3+}.[68,69] The introduction of strontium leads to the enhancement of clustering as well.[68]

4.3.1.4 Nickel-containing systems

The extrapolation of the isotherms χ_{Ni} (Fig. 4.1c) to the infinite dilution ($x \to 0$) gives the value of μ_{eff} = 3.4–3.86 μB for $LaNi_xGa_{1-x}O_3$. In the ESR spectra, a well-resolved signal from low-spin Ni^{3+} with $g = 2.157$ is observed, the signal from Ni^{2+} is absent. These data indicate that nickel at infinite dilution is in the spin state equilibrium. For $x = 0$, in the gallate doped with strontium some of Ni^{3+} remains in the spin equilibrium, like in the pure $LaGaO_3$, whereas some (~58%) is in the low-spin state. This is obviously related to Ni^{3+}, which happens to be in the vicinity of Sr^{2+}.[71]

The calculation of exchange interactions for the gallate doped only with nickel showed that the antiferromagnetic exchange is due to the formation of dimers Ni^{3+}(LS)–O–Ni^{3+}(LS) ($J = -20$ cm^{-1}), rather than Ni^{3+}(HS)–O–Ni^{3+}(LS) (Fig. 4.6a). When the gallate is doped with strontium, ferromagnetic dimers begin to play a major role in the exchange interactions Ni^{3+}(HS)–O–Ni^{3+}(LS) with $J = 10$ cm^{-1} (Fig. 4.6b).

Thus, in the case of nickel the doping with strontium causes an increase in clusterization, as well.

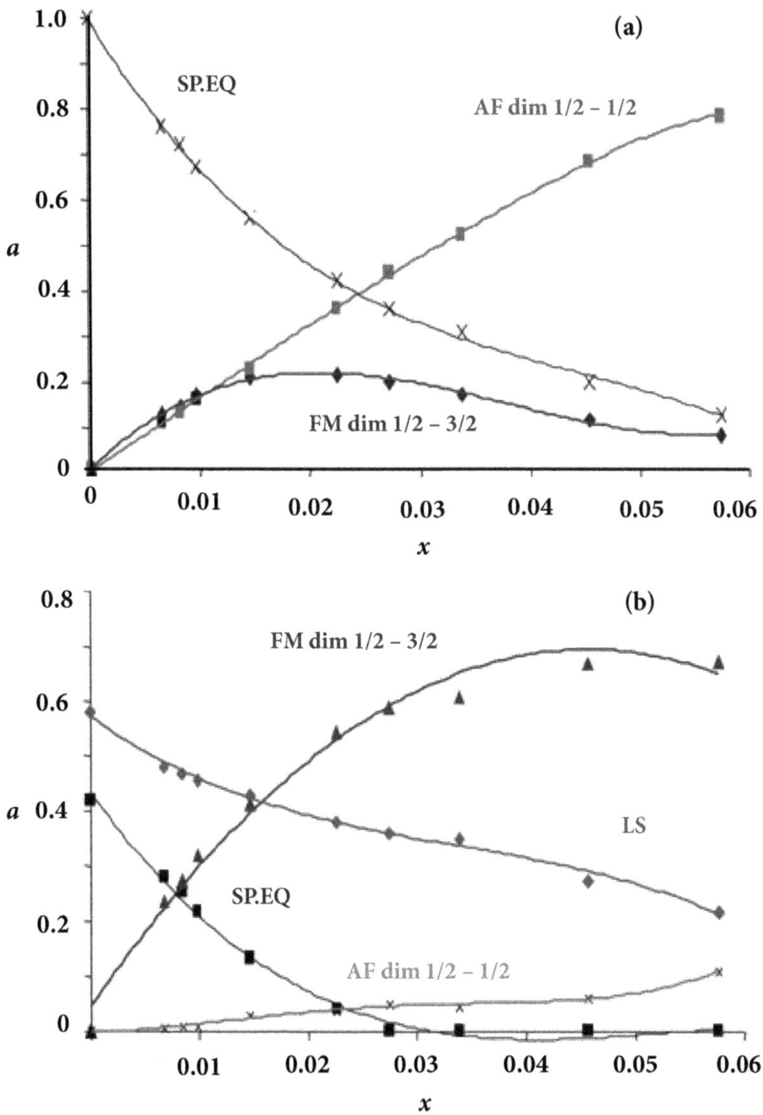

Figure 4.6 (a, b) Dependence of the aggregate contents on the concentration of nickel in solid solutions $LaNi_xGa_{1-x}O_3$ (a) and $La_{1-0.2x}Sr_{0.2x}Ni_xGa_{1-x}O_{3-\delta}$ (b).

4.4 Conductivity of Lanthanum Gallate Doped with Transition Element and Strontium in Ratio M:Sr = 5:1

The conductivity of the solid solutions in question increases as the temperature increases even in the region of low temperatures, which points to the semiconducting properties of the samples. The values of conductivity are low and depend on the content of a transition element. For the gallates doped only with a transition element, the dependence of the logarithm of specific electrical conductivity on the inverse temperature is linear and obeys Arrhenius equation.

Two sections can be selected in the temperature dependencies of solid solutions containing strontium, in Arrhenius equation being valid in each of them. Different values of the activation energy for these two linear sections point to a different character (mechanism) of conductivity (Fig. 4.7).

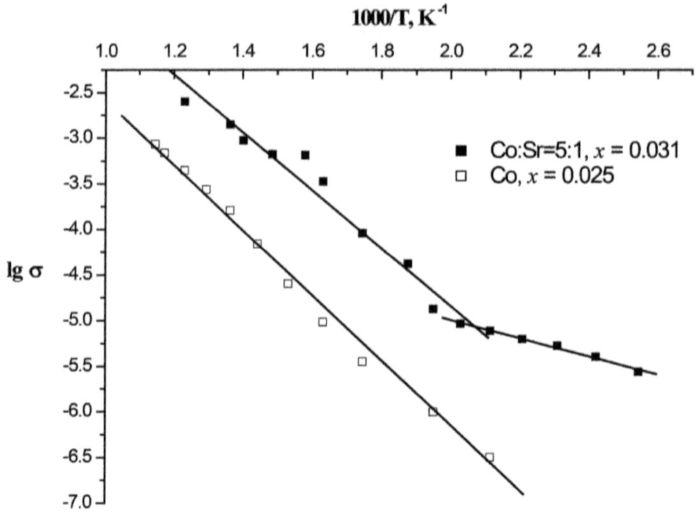

Figure 4.7 Dependencies of $lg\sigma$–$1/T$ for the systems with Co and Co:Sr = 5:1.

With the increase in temperature, ionic transition becomes more and more prominent. For the nickel-containing systems, the conductivity became fairly high ($lg\sigma > 0$) at $T > 500$ K.

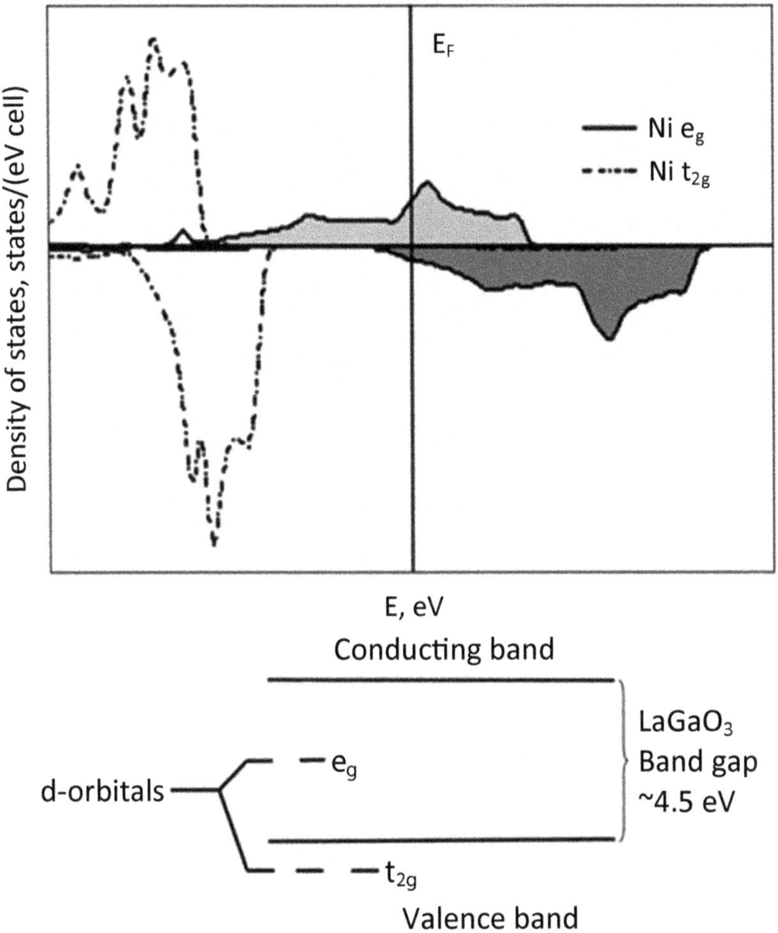

Figure 4.8 (a, b) Band levels for theoretical system $LaGa_{0.5}Ni_{0.5}O_3$ (a) and schematic representation of energy levels of *d*-orbitals and the band structure of $LaGaO_3$ (b).

To determine the nature of conductivity in a gallate doped with a transition element, band structure calculations (FLAPW GGA) for lanthanum gallate and lanthanum gallate in which one half of the gallium atoms was replaced with nickel ($LaGa_{0.5}Ni_{0.5}O_3$) were performed in the Institute of Solid State Chemistry of the Ural Branch of the Russian Academy of Sciences (Ekaterinburg). It was shown that the electronic conductivity in the solution appeared due to the fact that the band formed by the e_g-electrons of Ni^{3+} is within the forbidden band of the pure lanthanum gallate (Fig. 4.8a).[72] In diluted solid solutions, this band obviously degenerates into the impurity level (Fig. 4.8b).

Summarizing the investigation of magnetic properties of lanthanum gallates doped with strontium and a transition element with the ratio M:Sr = 5:1, the following conclusions can be drawn. First of all, the introduction of strontium, obviously, is not associated with an increase in the oxidation state of the transition element; however, it leads to the enhancement of clusterization, with the composition of clusters and the character of exchange greatly differing for the systems containing strontium and without strontium.

Hence, it can be stated that strontium participates in the formation of clusters. The replacement of lanthanum with strontium results in the appearance of the oxygen vacancy in its vicinity.

By all accounts, two atoms of the *d*-element happen to be near the vacancy and strontium, forming a "cluster," which stabilizes the structure.

The band structure calculation data for lanthanum gallate doped with nickel[72] allow us to extend them to all the elements studied and schematically represent the process of gallate doping. Figure 4.8a shows the zones of the wide band gap semiconductor $LaGaO_3$, the *d*-level of the transition metal, which is split in the octahedral field, is within the forbidden band (Fig. 4.8b). Chromium(III) (t_{2g}^3) forms three holes in the lanthanum gallate valence band, resulting in the appearance of a weak hole conductivity. Cobalt in gallate can be in two states. The low-spin cobalt with the configuration t_{2g}^6 has no free electrons. The high-spin cobalt ($t_{2g}^4e_g^2$) in dilute solutions makes insignificant contribution to the electronic conductivity. Consequently, in the gallate doped with Sr and Co a predominantly ionic conductivity may be expected, which was confirmed experimentally. Nickel in gallate, like cobalt, is in two states: low- and high-spin Ni^{3+}. One electron of low-spin Ni^{3+} ($t_{2g}^6e_g^1$) in the forbidden band of gallate accounts for the appearance of electronic conductivity. Manganese is entirely bound in very strong aggregates; therefore, relatively low conductivity can be expected in gallates doped with manganese. According to the data obtained (Fig. 4.9), the conductivity of gallate with manganese without strontium turned to be even higher than with strontium.

The data obtained allow us to recommend lanthanum gallate doped with strontium and nickel as a cathode for the solid oxide fuel cells due to the presence of, besides ionic, a considerable electronic conductivity. At the same time gallate doped with strontium and also with chromium and cobalt can serve as an electrolyte in the same cell, since their electronic conductivity is low, whereas the ionic (oxygen) conductivity increases in the sequence Mn < Cr < Ni < Co. Now two questions arise: (1) What happens as the fraction of strontium in the solid solutions increases? (2) Since in lanthanum gallate doped simultaneously with strontium and magnesium the ionic conductivity increases, what is the influence of magnesium or both strontium and magnesium on the electron structure of doped lanthanum gallate containing also a transition element?

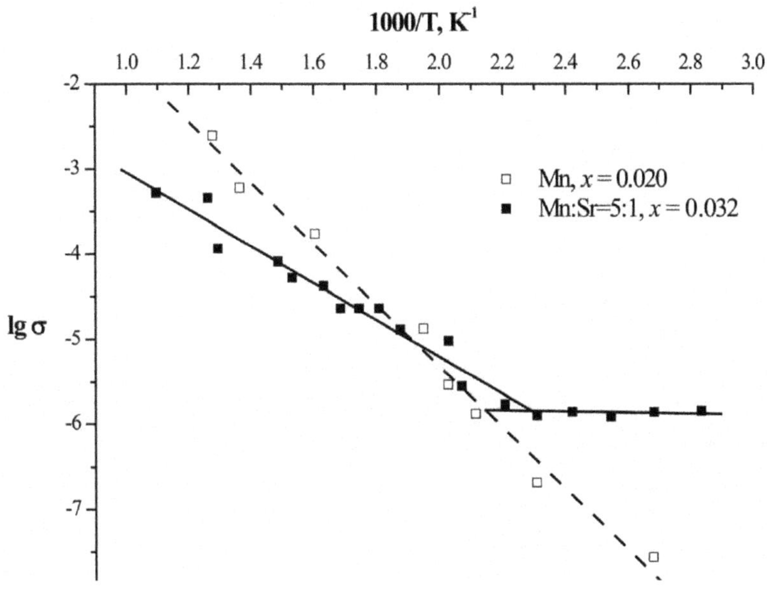

Figure 4.9 Dependencies $lg\sigma$–$1/T$ for $LaMn_xGa_{1-x}O_3$ and $La_{1-0.2x}Sr_{0.2x}Mn_xGa_{1-x}O_{3-\delta}$ systems.

4.5 The Impact of the Increased Fraction of Strontium and Introduction of Magnesium on Electron Structure and Electrophysical Properties of Doped Lanthanum Gallate

4.5.1 Chromium-Containing Systems

For the system with the ratio Cr:Sr = 5:1 the isotherms of magnetic susceptibility coincided with the isotherms of the system containing no strontium starting from x = 0.02,[73] and for the $La_{1-0.5x}Sr_{0.5x}Cr_xGa_{1-x}O_{3-\delta}$ solid solutions the isotherms lie substantially higher (Fig. 4.10), and at the infinite dilution the effective magnetic moment increases as the temperature increases (Fig. 4.11).

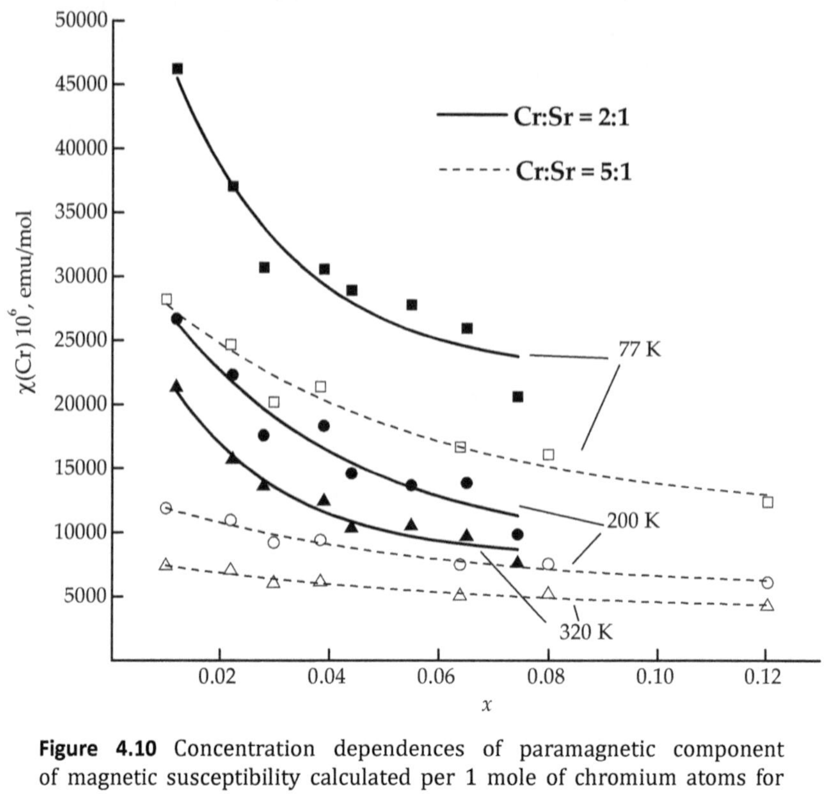

Figure 4.10 Concentration dependences of paramagnetic component of magnetic susceptibility calculated per 1 mole of chromium atoms for Cr:Sr = 5:1 and Cr:Sr = 2:1 systems at three temperatures.

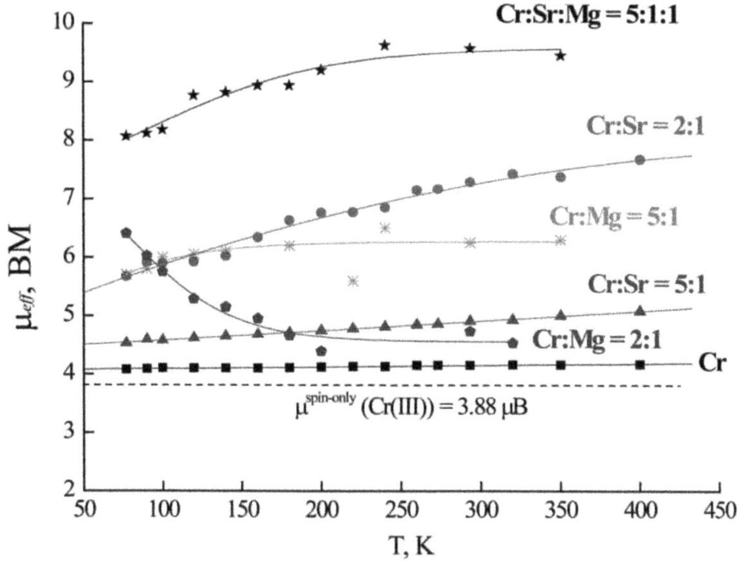

Figure 4.11 Temperature dependence of the effective magnetic moment at the infinite dilution for all chromium-containing systems.

At the infinite dilution the effective magnetic moments cannot be ascribed to the presence of single chromium atoms, no matter in which valence state it were: Cr^{2+}, d^4, ^5E_g, $\mu_{eff} \sim 4.90$ μB; Cr^{3+}, d^3, $^4A_{2g}$, $\mu_{eff} \sim 3.87$ μB; Cr^{4+}, d^2, $^3T_{1g}$, $\mu_{eff} \sim 2.83$ μB.

An assumption naturally arises about strontium atoms and vacancies in the oxygen sublattice associated with them taking part in the formation of clusters of paramagnetic atoms in the solid solution.

In this case such clusters appear to be so strong that they do not disintegrate even at the infinite dilution, as is commonly observed in the solid solutions of isomorphous substitution, in the $LaCr_xGa_{1-x}O_3$ system.[73]

In the doped $La_{1-0.2x}Sr_{0.2x}Cr_xGa_{1-x}O_{3-\delta}$ and $La_{1-0.5x}Sr_{0.5x}Cr_xGa_{1-x}O_{3-\delta}$ gallates we suggest the presence of single chromium atoms, antiferromagnetic Cr(III)–O–Cr(III) dimers, antiferromagnetic linear trimers, and certain large clusters with ferromagnetic exchange (clusters X), which are responsible for a nontypical temperature dependence of the effective magnetic moment at the infinite dilution (Fig. 4.11).[73]

The final equation for calculating the theoretical susceptibilities according to the approach of diluted solution will look like

$$\mu^2_{\text{experim}} = a_{clX}\mu^2_{clX} + a_{\text{dim}}\mu^2_{\text{dim}} + a_{\text{trim}}\,\mu^2_{\text{trim}} + (1 - a_{clX} - a_{\text{dim}} - a_{\text{trim}})\mu^2_{\text{mono}},$$

where μ_{mono} is the spin only value of the effective magnetic moment of chromium(III), and μ_{dim} and μ_{trim} were calculated for each temperature by HDVV model for two and three interacting atoms, respectively.[74]

In this manner, for an $S_1 = S_2 = 3/2$ couple with respect to $x = J_d/kT$ with $g_1 = g_2 = 2$, we obtain the following:

$$\mu^2_{\text{dim}} = \frac{g^2}{2}\frac{84e^{12x} + 30e^{6x} + 6e^{2x}}{7e^{12x} + 5e^{6x} + 3e^{2x} + 1}$$

The value of the exchange parameter $J_d = -12\ \text{cm}^{-1}$ was taken from ref. 70. For a homonuclear trimer Cr(III)-O-Cr(III)-O-Cr(III), we have one exchange parameter J_t. The spin Hamiltonian without regard to Zeeman interaction and its eigenvalues are

$$\hat{H} = -2J_t(\hat{S}_1\hat{S}_2 + \hat{S}_1\hat{S}_3);$$

$$E_t(J_t, S) = -J_t[S(S+1)] - S_{23}(S_{23}+1) - S_1(S_1+1),$$

where, according to the rules of the moment addition the total (S) and intermediate (S_{23}) spin take the values

$$S = S_1 + S_{23}, S_1 + S_{23} - 1, \ldots |S_1 - S_{23}|;$$

$$S_{23} = S_2 + S_3, S_2 + S_3 - 1, \ldots |S_2 - S_3|.$$

Using Van Vleck's equation it is easy to obtain the final expression for the magnetic moment of three exchange bonded atoms $S_1 = S_2 = S_3 = 3/2$, with $g_1 = g_2 = g_3 = 2, y = J_t/kT$:

$$\mu^2_{\text{trim}} = \frac{g^2}{3}\frac{247{,}5e^{21y} + 126e^{18y} + 52{,}5e^{15y} + 141e^{12y} + 52{,}5e^{11y} + 15e^{10y} + 1{,}5e^{7y} + 15e^{6y} + 52{,}5e^{5y} + 1{,}5e^{3y} + 15}{10e^{21y} + 8e^{18y} + 6e^{15y} + 12e^{12y} + 6e^{11y} + 4e^{10y} + 2e^{7y} + 4e^{6y} + 6e^{5y} + 2e^{3y} + 4}$$

The only difference from the case of dimers is in the fact that in the equation for the susceptibility we must sum up by the intermediate spin with the aim of taking into account the multiplicity of degeneration of the spin multiplates with the same values of the total spin.

Our calculations gave us the optimal value of the exchange parameter for the trimers $J_t = -20$ cm^{-1}.

Having neither the data on the structure of clusters X nor the possibility to describe the dependence $\mu_{eff} = f(T)$ (Fig. 4.11), with the help of HDVV model, when determining μ_{clX} we postulated that the effective magnetic moment at the infinite dilution for the system with the ratio of Cr:Sr = 2:1 is determined by clusters X only, i.e., their fraction is 1.

Then, as the calculation showed, the effective magnetic moments for the system with Cr:Sr = 5:1 is the superposition of 20% of clusters and 80% of monomers, which agrees with the difference in the ratio Cr:Sr for the systems under study. The data of calculation are given in Fig. 4.12a,b. The difference between theoretical and experimental magnetic susceptibilities for both systems does not exceed 2% (Fig. 4.13a,b).

We tried also to exclude one or other type of clusters—monomers, antiferromagnetic dimers or trimers, but this resulted in a substantial discrepancy between the data of theory and experiment (up to 7%), hence we concluded that the model of the structure of the solid solution we advanced is the most plausible.[75]

Attention is drawn to some special features of the changes in the cluster fractions. For the solutions containing no strontium we have a typical pattern of antiferromagnetic dilution—as the concentration of a paramagnetic component increases, the fraction of monomers decreases monotonously, the fraction of antiferromagnetic dimers and trimers increases. As we introduce strontium in the ratio Cr:Sr = 5:1 the clusters X with competing antiferro- and ferromagnetic exchange appear along with monomers and dimers. Their fraction decreases as the chromium concentration in the solution increases (the same is true for monomers) (Fig. 4.12a). An increase in the content of strontium to the ratio Cr:Sr = 2:1 results in a relative increase in the fraction of clusters X and in a nonmonotonous change in the fraction of monomers in the solid solution (maximum at $x \sim 0.045$). As in two previous cases, as the concentration of chromium in the solution increases, the fraction of dimers and trimers increases (Fig. 4.12b). Such a trend has a sufficiently simple explanation.

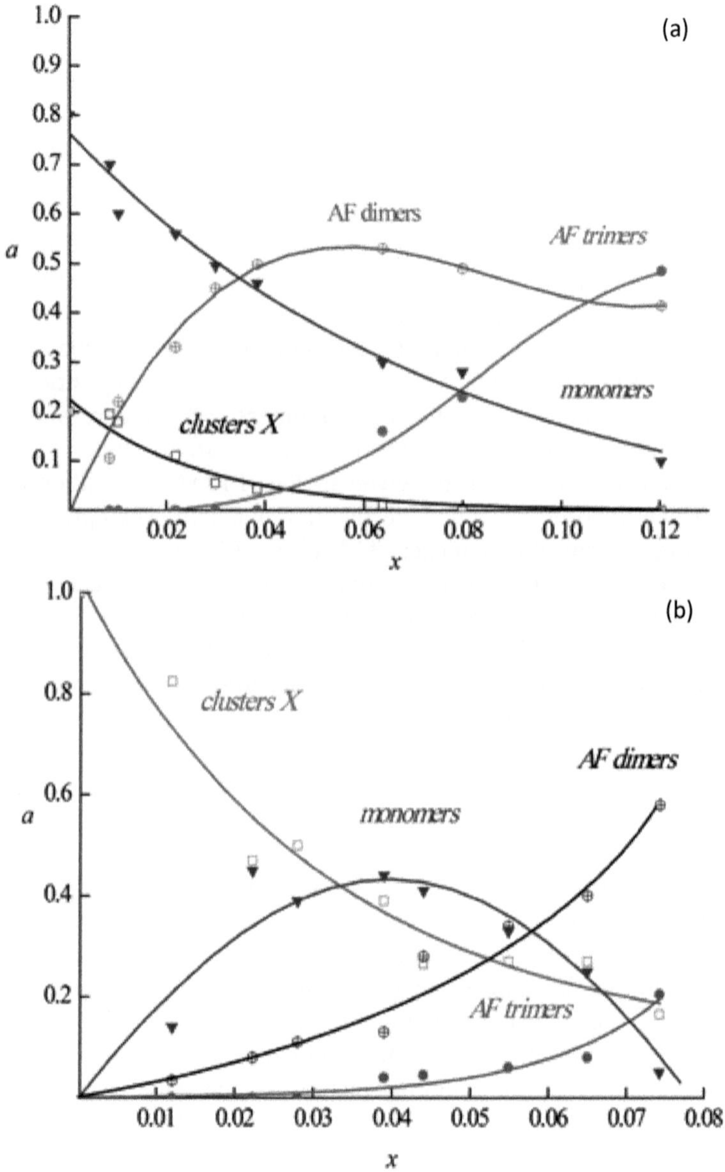

Figure 4.12 (a, b) Plots of the fractions of clusters and single atoms vs. chromium concentration for the systems with various strontium content—$La_{1-0.2x}Sr_{0.2x}Cr_xGa_{1-x}O_{3-\delta}$ (*a*), $La_{1-0.5x}Sr_{0.5x}Cr_xGa_{1-x}O_{3-\delta}$ (b).

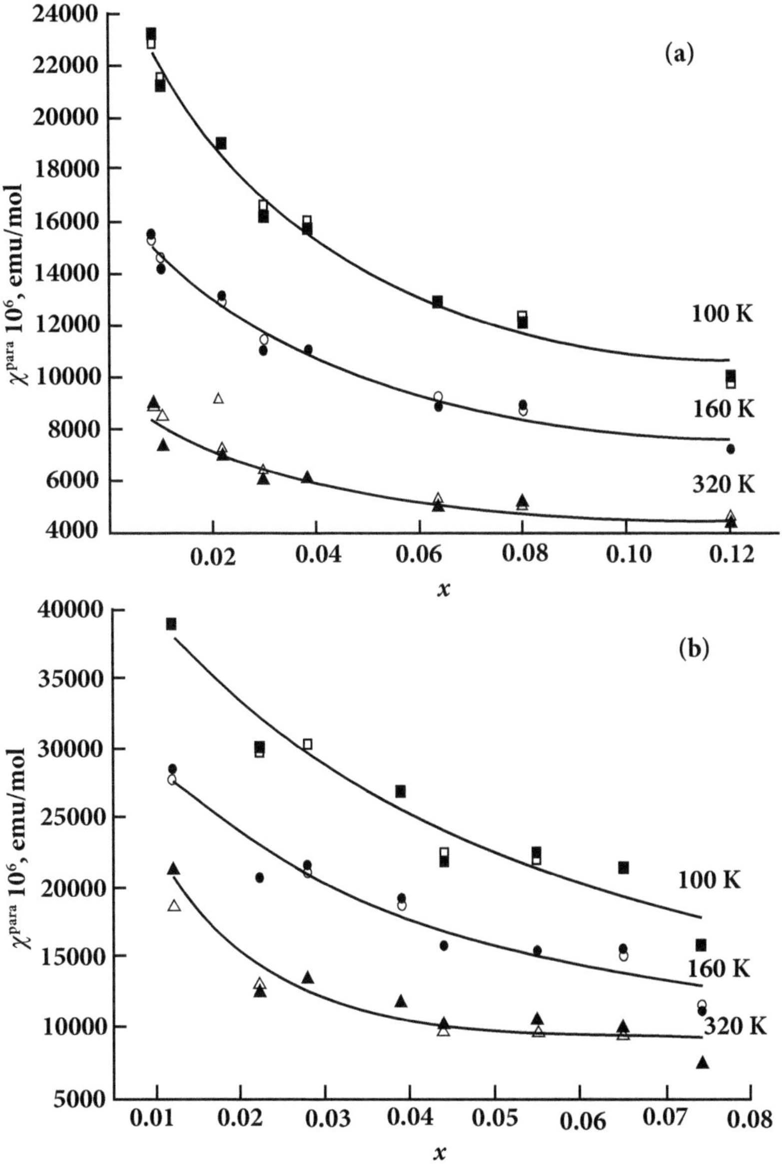

Figure 4.13 (a, b) Calculated (open symbol) and experimental (closed symbol) paramagnetic components of magnetic susceptibility: $La_{1-0.2x}Sr_{0.2x}Cr_xGa_{1-x}O_{3-\delta}$ (a) and $La_{1-0.5x}Sr_{0.5x}Cr_xGa_{1-x}O_{3-\delta}$ (b) for three temperatures.

(1) *The interval* $0 < x < 0.045$. If at the infinite dilution there remain only ferromagnetic clusters (Fig. 4.12b), taking into account the ratio Cr:Sr = 2:1 there will be a deficit of strontium atoms to form a real oxygen vacancy. Hence a "half" of a vacancy remains in the oxygen site, i.e., it must be an electron:

$$\mathrm{SrO} + \mathrm{La}_{\mathrm{La}}^{\times} + 1/2\mathrm{O}_{\mathrm{O}}^{\times} \rightarrow \mathrm{Sr}_{\mathrm{La}}' + 1/2V_{\mathrm{O}}^{\bullet\bullet} + 1/2\mathrm{La}_2\mathrm{O}_3,$$

which can exist either in a couple with a "hole" $h^+ \leftrightarrow \bar{e}$ of the type of Wannier–Mott exciton. As the quantity of strontium increases, the situation arises, when two strontium atoms are in the close vicinity, which results in the formation of an oxygen vacancy:

$$2\mathrm{SrO} + 2\mathrm{La}_{\mathrm{La}}^{\times} + \mathrm{O}_{\mathrm{O}}^{\times} \rightarrow 2\mathrm{Sr}_{\mathrm{La}}' + V_{\mathrm{O}}^{\bullet\bullet} + \mathrm{La}_2\mathrm{O}_3.$$

In this case a vacancy located between chromium atoms disrupts the integrity of the superexchange channel, which results in one of the chromium atoms ceases to take part in the exchange and behaves as a monomer. At the same time this decreases the relative number of clusters X. The fraction of dimers is insignificant in this region and comprises ~10% for $x \sim 0.045$.

(2) *The interval* $0.045 < x < 0.075$. At the critical concentration ($x \sim 0.045$), when the fraction of monomers is maximal, the conditions are created for their further unimpeded aggregation not only into dimers, but also into trimers, their number monotonously increasing starting from $x \sim 0.045$. At this concentration the influence of strontium (by the formation of oxygen vacancies) becomes minimal, since the fraction of clusters X is small and remains almost constant up to $x = 0.075$.

As the result the fraction of antiferromagnetic dimers and trimers abruptly increases with regular decrease in the fraction of monomers, just as in the case of diluting typical antiferromagnets.

It is interesting that for the $\mathrm{La}_{1-0.5x}\mathrm{Sr}_{0.5x}\mathrm{Cr}_x\mathrm{Ga}_{1-x}\mathrm{O}_{3-\delta}$ solid solution the deviations from Curie–Weiss law are observed, which are typical for ferrimagnets. The bend in the plot of $1/\chi_{Cr}$ vs. T shows itself in the region of low temperatures and becomes greater as the concentration of chromium increases (Fig. 4.14).

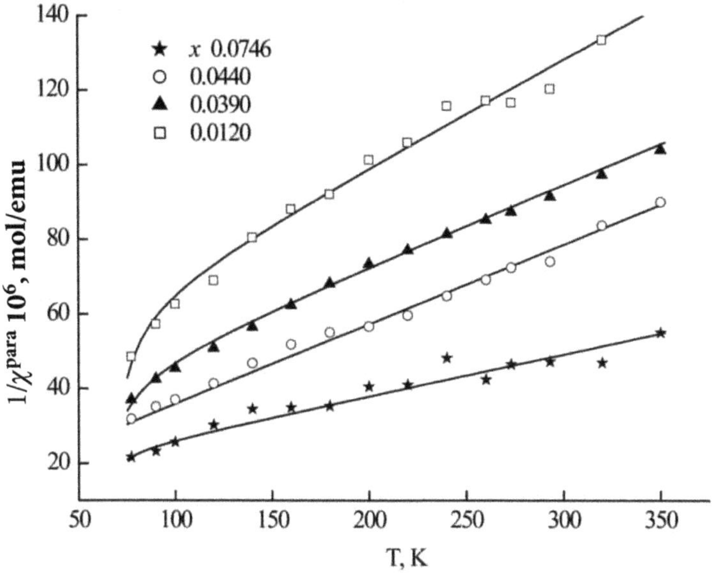

Figure 4.14 Plot of $1/\chi - T$ for $La_{1-0.5x}Sr_{0.5x}Cr_xGa_{1-x}O_{3-\delta}$.

The calculations of magnetization also point to the presence of highly nuclear clusters, their behavior having much in common with the behavior of super paramagnetic particles.

The following requirements arise from the determination of super paramagnetic state for the experimental corroboration of its presence in the system under study: (1) the absence of hysteresis in the magnetization curve and (2) the superposition of magnetization curves obtained at various temperatures and plotted as $M-H/T$ (Fig. 4.15).[76-78] The second requirement is not extended to the cases, when it appears impossible to neglect the interactions between superparamagnetic particles. In our case for diluted solutions we can state with certainty that there can be no interactions between such particles.

We can examine the character of magnetization at low temperatures with the help of the plots $H/M-M^2$ we considered Belov–Arrott plots (Fig. 4.16).[79]

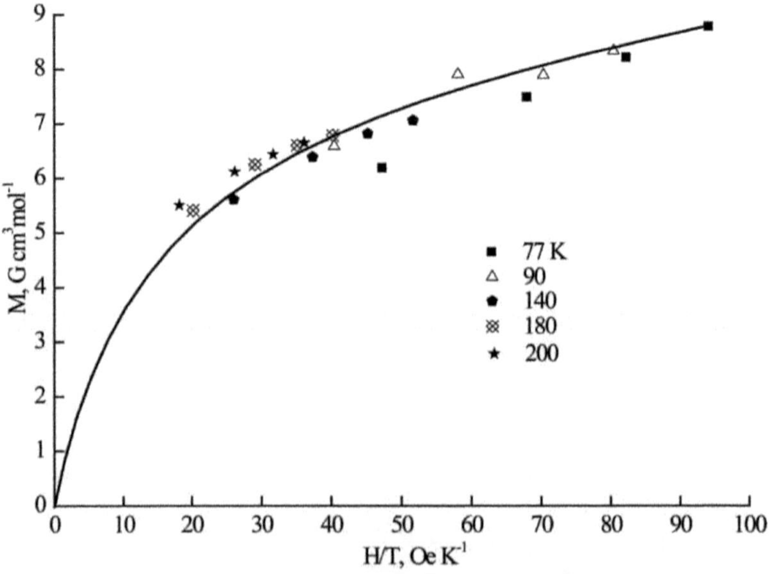

Figure 4.15 Superposition of magnetization curves for the Cr:Sr = 2:1 system $x = 0.0223$ at various temperatures and magnetic field strength.

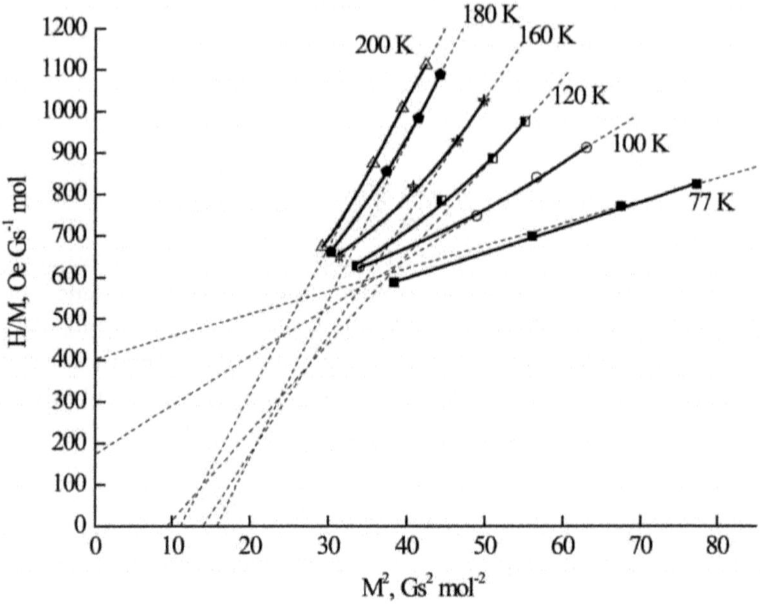

Figure 4.16 Belov–Arrott plots for the Cr:Sr = 2:1 solid solution, $x = 0.0223$.

The plots $H/M = f(M^2)$ must be linear and crossing the axis of ordinates at positive α for $T > T_c$ and at negative α for $T < T_c$. Zero value of the thermodynamic coefficient (α) corresponds to the temperature of magnetic ordering. In our case for all the compositions $\alpha > 0$ at low temperatures, which unambiguously points to the absence of spontaneous magnetization and, consequently, of long order magnetic order in the solid solutions.

Using the concept of superparamagnetism[80] we can estimate the average sizes of clusters and describe their magnetic properties.[81,82]

In the small fields just like for atomic paramagnetism we can write the equation for Langevin's function, which in a limiting case will turn into Curie–Weiss by equation (*):

$$\chi = M/H = M_0 M_s/3k(T - \omega M_0 M_s/3k) = C/(T - \Theta'), \qquad (*)$$

where M_0 is the magnetization of a cluster is determined by Eq. (**):

$$M_0 = I_0 \int v G(v) dv = I_0 v_{av}, \qquad (**)$$

where v_{av} is an average size of the cluster, and ω the constant of intercluster interaction.

According to Equations (*) and (**) the calculation formula for the determination of the sizes of ordered micro regions is the following:

$$r = \left\{ \frac{9kCM}{4\pi M_S I_o \rho} \right\}^{1/3},$$

where M is the molecular weight, and ρ the sample density. We used for our calculation the theoretical density taken as 4.23 g/cm^3.

We also used in our calculations a theoretical value of magnetization of saturation of a cluster substance I_0 = 16700 Gs · cm^3 · mol^{-1}, which was determined by next equation:

$$I_0 = n\mu_B N_A,$$

where n is the number of electrons on a paramagnetic centre, μ_B the Bohr magneton, and N_A the Avogadro number.

This estimation of the sizes of highly nuclear clusters is, of course, sufficiently rough, but it gives $r \approx 1–2$ nm.

A similar tendency, i.e., the dependence of susceptibility on magnetic field is observed for the whole series of the solid solutions. The obtained dependencies are similar for the solid solutions with various concentrations; hence, we give them for one concentration only. In all the diluted solutions the hysteresis loop is absent. For more concentrated solutions coercive forces are small and do not exceed 0.12 $Gs \cdot cm^3 \cdot g^{-1}$. All these facts allow us to consider the clusters formed in the system with Cr:Sr = 2:1 as superparamagnetic.

As we add to strontium as a doping element magnesium with the ratio Cr:Sr:Mg = 5:1:1 $La_{1-0.2x}Sr_{0.2x}Cr_xGa_{1-1.2x}Mg_{0.2x}O_{3-\delta}$, the susceptibility isotherms appear to lie even higher than for the system with Cr:Sr = 2:1 described above (Fig. 4.17). Taking into account the fact that Mg occupies the sites of Cr and Ga rather than lanthanum as strontium does we had to suggest that other clusters with even stronger ferromagnetic exchange are formed in this system—clusters Y presumably containing the atoms of Cr, Mg, and accompanying vacancies.

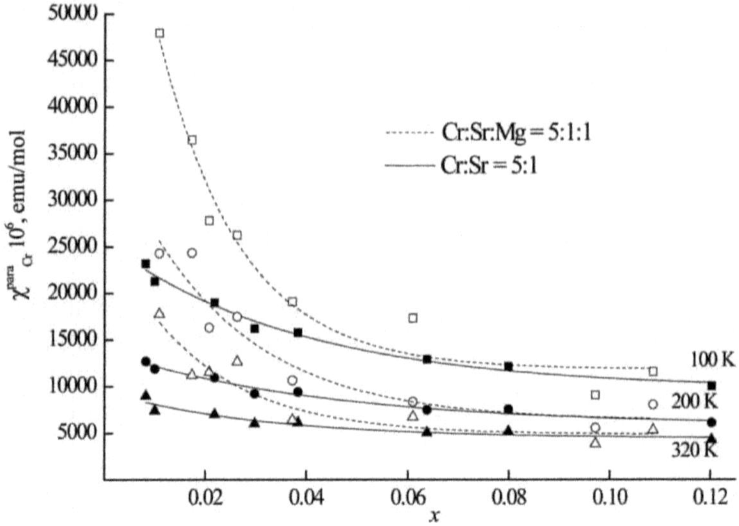

Figure 4.17 Concentration dependences of paramagnetic component of magnetic susceptibility calculated per 1 mole of chromium atoms for Cr:Sr:Mg = 5:1:1 and Cr:Sr = 2:1 systems at three temperatures.

In our HDVV calculations we took into consideration single chromium atoms, antiferromagnetic dimers, antiferromagnetic linear trimers, clusters X,[74] and a new type of clusters—clusters Y. The influence of magnesium on the number of various clusters of chromium atoms can be traced upon comparing the concentration dependences of the cluster fractions for Cr:Sr:Mg = 5:1:1 systems (Fig. 4.18).

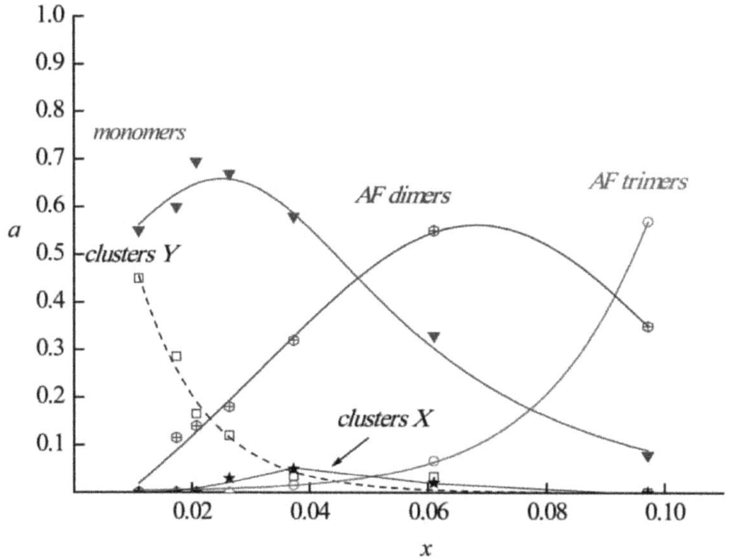

Figure 4.18 Plots of the fractions of clusters and single atoms vs. chromium concentration for the $La_{1-0.2x}Sr_{0.2x}Cr_xMg_{0.2x}Ga_{1-1.2x}O_{3-\delta}$ system.

The special features of the changes in the cluster fractions depend both on qualitative and quantitative composition of the system. For the solutions containing neither strontium nor magnesium, $LaCr_xGa_{1-x}O_3$, we have a typical pattern of the dilution of antiferromagnets—as the concentration of paramagnetic component increases, the fraction of monomers decreases, the fraction of antiferromagnetic dimers increases.[70]

The introduction of chromium and strontium into lanthanum gallate with the ratio Cr:Sr = 5:1 results in clusters X apart from monomers, dimers, and increasing fraction of trimers. These clusters X have no less than 4 paramagnetic atoms, are somehow associated with the vacancies in the sublattice of oxygen, and have a distinct ferromagnetic component in the exchange. The introduction of magnesium in addition to strontium results in some other clusters—clusters Y even greater than clusters X. Both types of clusters are preserved at the infinite dilution, but their fraction decreases as the concentration increases. This points again to the fact that strontium atoms and vacancies associated with them are included into these clusters with a strong bonding. We can assume that as concentration increases, the additional chromium atoms become located near the vacancies thus behaving like monomers, their fraction increasing as the concentration increases. Then the dimer and trimer clusters appear with antiferromagnetic exchange.

For the systems containing only magnesium as a diamagnetic doping element we also find the field-dependent susceptibility over the whole temperature range (Fig. 4.19). And again the magnetization vs. H/T curves and positive coefficients α point to the absence of spontaneous magnetization, thus suggesting super paramagnetic behavior of the systems with Cr:Mg = 5:1 and Cr:Mg = 2:1. An estimation of the sizes of clusters (clusters Y) give $r \sim$ 1.1–1.5 nm. The values of μ_{eff} and their temperature dependencies at the infinite dilution (Fig. 4.11) point to the presence of clusters and the absence of chromium atoms in a higher oxidation state. Magnesium as well as strontium may give rise to oxygen vacancies and electrons at the sites of vacancies.[83]

$$\text{MgO} + \text{Ga}_{\text{Ga}}^{\times} + 1/2\text{O}_{\text{O}}^{\times} \rightarrow \text{Mg}_{\text{Ga}}^{\prime} + 1/2V_{\text{O}}^{\bullet\bullet} + 1/2\text{Ga}_2\text{O}_3,\ \ 1/2V_{\text{O}}^{\bullet\bullet} \equiv e$$

Migration of electrons from one paramagnetic site to another must result in a strong ferromagnetism, which we observe at the infinite dilution.

The special features of the systems under study are directly associated with both qualitative and quantitative composition. A variation of one parameter only (concentration or the nature of the substituent) results in irreversible changes in the system as a whole, which are associated with complex and manifold interactions between separate atoms and their aggregates.

Let us see how this concept works in other transition element-containing systems.

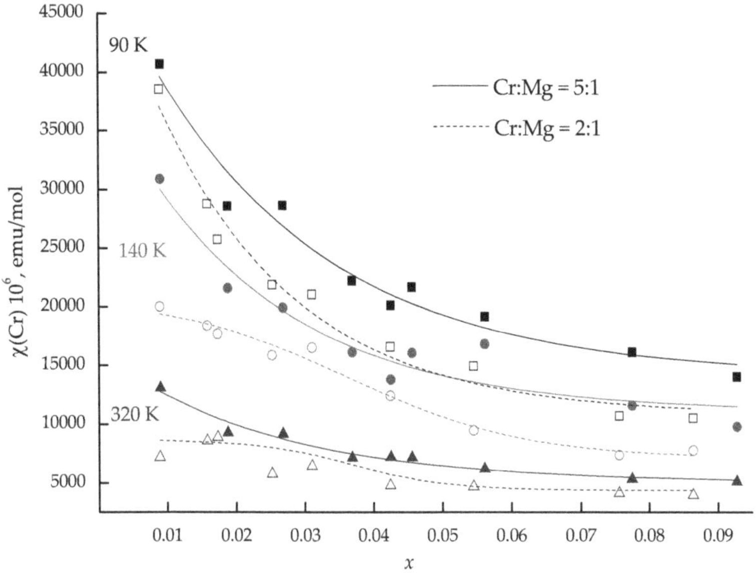

Figure 4.19 Plots of magnetic susceptibility vs. x for the $LaCr_xGa_{1-1.2x}Mg_{0.2x}O_{3-\delta}$ (Cr:Mg = 5:1) and $LaCr_xGa_{1-1.5x}Mg_{0.5x}O_{3-\delta}$ (Cr:Sr = 2:1) systems.

4.5.2 Nickel-Containing Systems

4.5.2.1 $La_{1-0.5x}Sr_{0.5x}Ni_xGa_{1-x}O_{3-\delta}$, $LaNi_xGa_{1-1.2x}Mg_{0.2x}O_{3-\delta}$ and $LaNi_xGa_{1-1.5x}Mg_{0.5x}O_{3-\delta}$ systems

Magnetochemical behavior of these systems has a lot in common; hence, we shall consider them in one paragraph. A similarity in the behavior shows itself in a similar character of the isotherms of paramagnetic component of magnetic susceptibility. For the systems with the ratios Ni:Sr = 2:1, Ni:Mg = 5:1, and Ni:Mg = 2:1 the susceptibility monotonously increases in the region of small nickel concentrations and is almost constant in a more concentrated region (Figs. 4.20, 4.21).

A comparison between these systems and the system with Ni:Sr = 5:1 shows two substantial differences: First, in the systems with magnesium and with greater quantity of strontium local extremes are absent over the whole concentration interval, whereas in the system with Ni:Sr = 5:1 a distinct maximum is observed at $x \sim 0.045$, second, the susceptibilities are essentially smaller than for Ni:Sr = 5:1.

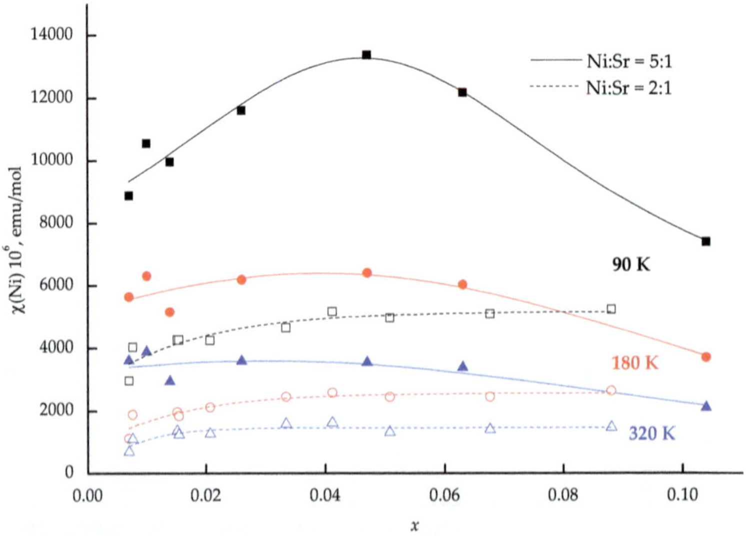

Figure 4.20 Plots of paramagnetic components of magnetic susceptibility vs. nickel content for the systems with Ni:Sr = 5:1 and Ni:Sr = 2:1 at three temperatures.

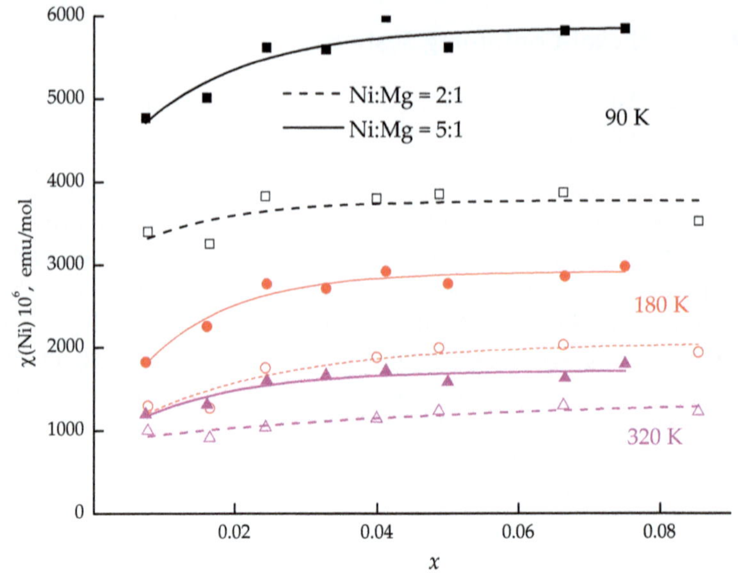

Figure 4.21 Plots of paramagnetic components of magnetic susceptibility vs. nickel content for the systems with Ni:Mg = 5:1 and Ni:Mg = 2:1 at three temperatures.

As has been shown earlier,[71] for the system with Ni:Sr = 5:1, the effective magnetic moments at the infinite dilution were described as a superposition of effective magnetic moments of low-spin nickel(III) and of nickel(III) in the state of spin equilibrium.

On the basis of the ideas about mutual influence of dia- and paramagnetic atoms in the Ni:Sr = 2:1, Ni:Mg = 5:1, and Ni:Mg = 2:1 systems we can expect some shift in the spin equilibrium resulting in an increase in the quantity of low-spin nickel(III).

We emphasize that for the Ni:Mg = 5:1 system such a tendency is typical owing to magnesium atoms being located in gallium sites of perovskite structure in distinction to strontium, therefore its depolarizing effect on nickel atoms must be stronger.

Let us try to make sense of such an unconventional behavior of these three systems based on lanthanum gallate examining the possible cases of interatomic interactions and various valence states of nickel.

(1) *The formation of Ni^{4+} in the quantity equivalent to the quantity of introduced heterovalent substituent.*

Such a situation is possible taking into account the opinions published by this moment in literature.[27,55,84,85]

Looking at Fig. 4.22 we see that the values of μ_{eff} at the infinite dilution are much lower than 1.8 μB typical for low-spin isolated atoms of Ni^{3+}.

Assuming that these low values are associated with some fraction of diamagnetic particles (Ni^{4+} must be low spin), we find that the fraction of Ni^{4+} in all these three systems varies between 50% and 70%, which is impossible in terms of the mole ration of nickel and a substituent.

(2) *Disproportioning of Ni^{3+}*. We can address ourselves to the process of disproportionation resulting in two forms—Ni^{4+} and Ni^{2+}, for Ni^{2+} the diamagnetic state being not excluded in the square planar surrounding, when oxygen vacancies are in axial position.

The idea of disproportioning is interesting since, at first sight, it can adequately account for the local maxima in the plots $1/\chi_{Ni} - T$, which appear at very low concentrations of the dopant (Fig. 4.23a,b,c).

For all three cases we have two anomalies in the temperature dependencies in the region of $T \sim 180$ and 300 K. One of them, about 180 K may be accounted for by the transfer $Ni^{2+}_{dia} \rightarrow Ni^{2+}_{para}$.

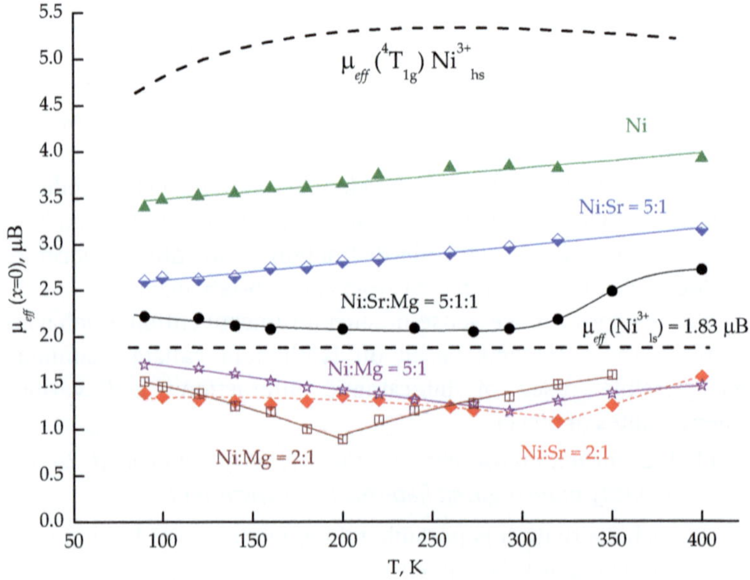

Figure 4.22 Temperature dependence of the effective magnetic moments at infinite dilution for all the nickel-containing systems.

However, an attempt to represent the effective magnetic moments at the infinite dilution as a superposition of the moments of nickel (II, III, IV) and possible clusters of the type Ni^{3+}(LS)-Ni^{2+} does not agree with the experimental data. In this case it is worthwhile to note that in the EPR spectra only one wide line was observed attributed to the clusters of low-spin Ni^{3+}.

Possible location of nickel and doping diamagnetic atoms in the case of nickel disproportionation in the solid solutions.

We must bear in mind the fact that for more concentrated samples the anomalies also exist, but their absence in the plots must be associated with an increase in the fractions of low dimension clusters (as it was in chromium-containing systems), which results in averaging of the susceptibility.

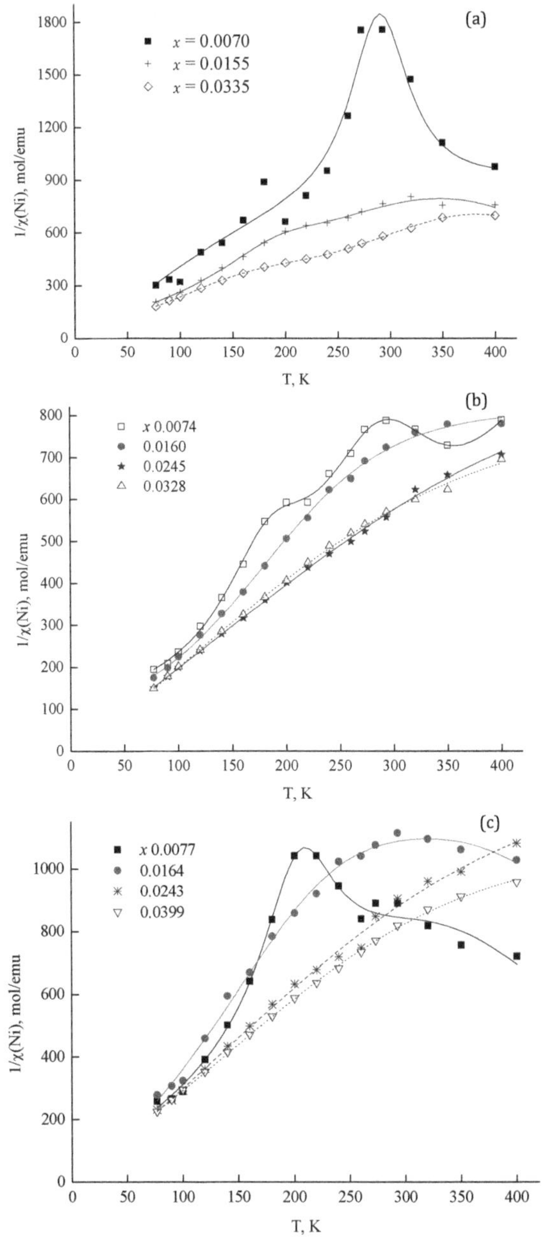

Figure 4.23 (a, b, c) Plots of inverse susceptibility vs. temperature for various systems and concentrations of the solid solutions: (a) Ni:Sr = 2:1; (b) Ni:Mg = 5:1; (c) Ni:Mg = 2:1.

This structural transition from rhombohedral to orthorhombic syngony[86] occurs as the temperature increases and can result in an additional low-spin nickel atoms appearing in the solution or in an increase in the crystal field splitting.

(3) *Formation of high-nuclearity clusters.* To develop our ideas about this topic, we must recall the situation with chromium-containing systems, where we observed an increase in clustering of paramagnetic atoms as the concentration of heterovalent diamagnetic substituents in lanthanum gallate increased. Clustering is attested by nontypical dependencies of effective magnetic moment at the infinite dilution, which cannot be described in terms of the theory of isotropic exchange of Heisenberg–Dirac–Van Vleck.

In this case, judging from the absolute values of magnetic moment (Fig. 4.11) in all the compositions the ferromagnetic component of the exchange exceeds the antiferromagnetic component, and we observe an abrupt increase in μ_{eff}. This may be accounted for by the fact that the distortions of the surrounding of chromium atoms resulting from the presence of vacancies of heterovalent dopants decreases the overlapping between t_{2g}-orbitals of Cr^{3+} and p-orbitals of oxygen, thus decreasing the antiferromagnetic component in the exchange. Ferromagnetic exchange is not associated with orbital overlapping.

As opposed to Cr^{3+}, Ni^{3+} in any spin state has e_g-electrons. It is just these d_{z2} and d_{x2-y2} orbitals directly overlap with p-orbitals of oxygen atoms, and as a consequence, a strong antiferromagnetic exchange results.

Therefore, we can suggest with certainty that at the infinite dilution for nickel-containing systems, where the effective magnetic moment is much lower than 1.83 μB (Fig. 4.22) and the character of their temperature dependence is also nontypical, there are clusters of high nuclearity with competing antiferro- and ferromagnetic exchange, but antiferromagnetic component appears to prevail, thus substantially decreasing the effective magnetic moments.

4.5.2.2 $La_{1-0.2x}Sr_{0.2x}Ni_xMg_{0.2x}Ga_{1-1.2x}O_{3-\delta}$ system

This system differs from all the nickel-containing solid solutions. From temperature dependence of effective magnetic moment at the infinite dilution we can see that the character of the dependence is nonmonotonous. There is a moderate decrease in μ_{eff} from 2.4 to 2.1 μB in the temperature range 77–300 K and an abrupt increase after 300 K.

In this system the effective magnetic moments seem to be well described by the ideas about the existence of single Ni^{3+} atoms. However, the situation is more complicated. We deal with clusters of high nuclearity, but in distinction to the systems with Ni:Sr = 2:1, Ni:Mg = 5:1, Ni:Mg = 2:1 upon simultaneous introduction of strontium and magnesium ferromagnetic exchange channels begin to play greater role thus increasing the effective magnetic moment. A similar situation was observed for the Cr:Sr:Mg = 5:1:1 system, where we observed the greatest magnetic moments of all the systems under study.

By and large, as the concentration of nickel increases, the greater number of antiferromagnetic dimers of low-spin nickel are formed, which determines the shape of isotherms in Fig. 4.24.

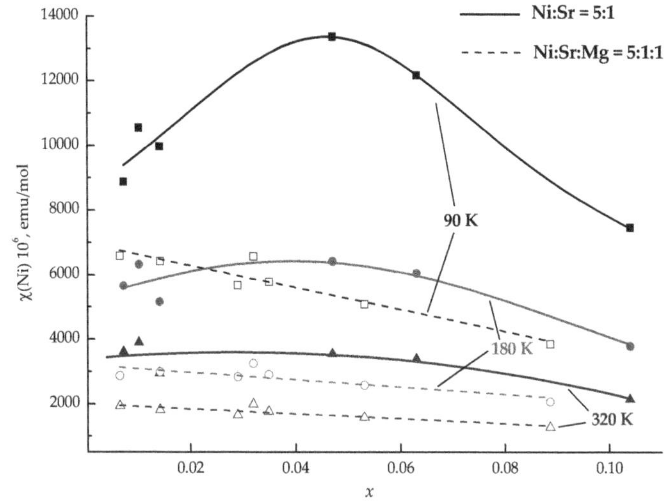

Figure 4.24 Concentration dependence of paramagnetic component of magnetic susceptibility for the Ni:Sr = 5:1 and Ni:Sr:Mg = 5:1:1 systems at three temperatures.

We observe no local maxima in the dependence of inverse susceptibility on temperature, only the deviations from Curie–Weiss law at high temperatures typical for temperature-independent paramagnetism (Fig. 4.25).

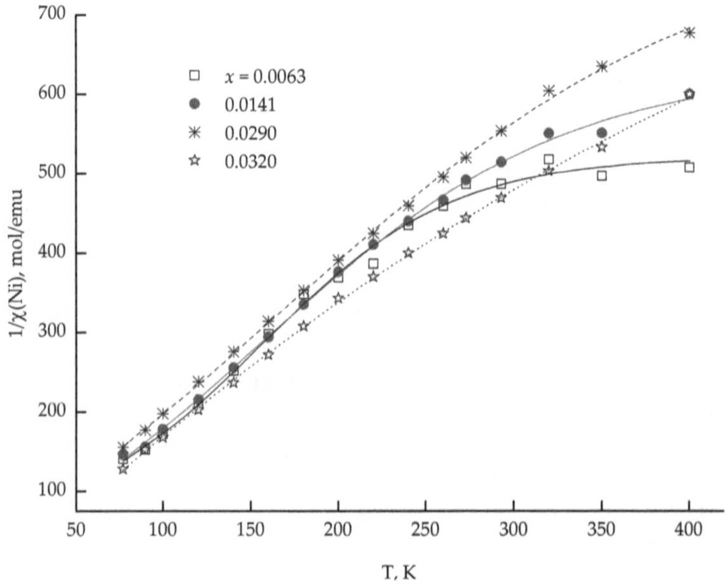

Figure 4.25 Plots of inverse susceptibility vs. temperature for $La_{1-0.2x}Sr_{0.2x}Ni_xGa_{1-1.2x}Mg_{0.2x}O_{3-\delta}$ for various concentrations (in the insert for $x = 0.0349; 0.0531; 0.0887$).

4.5.3 $LaCo_xMg_{0.2x}Ga_{1-1.2x}O_{3-\delta}$ and $LaCo_xMg_{0.5x}Ga_{1-1.5x}O_{3-\delta}$ Systems

4.5.3.1 X-ray diffraction and phase composition

The powders of sintered samples at 1723 K, were identified as single-phase with structure of orthorhombic $LaGaO_3$ (S.G. *Pbnm*) and rhombohedral $LaGaO_3$ (S.G. *R*-3*c*), which resulted from Rietveld refinement (the profile of the X-ray pattern are given in Fig. 4.26). As can be seen in Fig. 4.26 there is a line in the region of $2\theta \sim 29°$, which is a K_β line. It appears due to the special features of the experiment upon using Ni filter; a shoulder in the most intensive line at $2\theta \sim 32°$ is an edge of the absorption line of Ni filter. K_β lines were taken into account automatically on treating the patterns in TOPAS software.

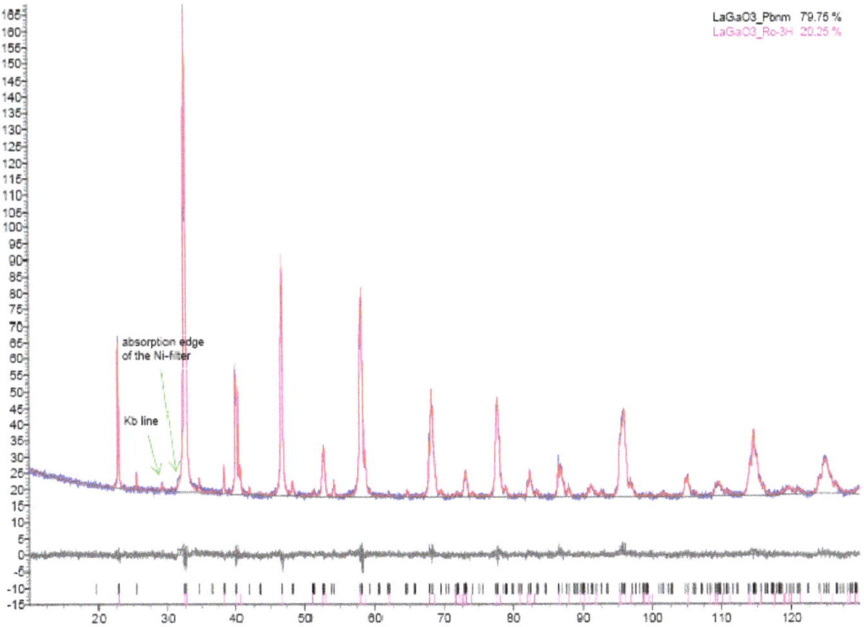

Figure 4.26 Result of the Rietveld refinement for the pattern recorded at room temperature for the system $LaGa_{1-1.2x}Co_xMg_{0.2x}O_3$, x(Co) = 0.0845, y(Mg) = 0.0197. R_{wp} = 5.02% R_p = 3.95%, GOF = 1.23, R-Bragg = 1.74.

It is interesting that we were the first to observe the coexistence of two phases of $LaGaO_3$ with different symmetry. In Fig. 4.27 the calculation of the pattern from Fig. 4.26 by Rietveld method is shown, which was carried out using *Pbnm*, *R*-3*c*, and upon their joint presence. It is seen that the best results are obtained for the last case.[87]

For both systems Co:Mg = 5:1 and Co:Mg = 2:1 the solid solutions with low concentrations of doping elements have orthorhombic structure. The mass fraction of rhombohedral phase increases nonlinearly as the quantity x(Co) + y(Mg) increases (Fig. 4.28). It is seen also that for the system with lower quantity of magnesium $LaCo_xMg_{0.2x}Ga_{1-1.2x}O_{3-\delta}$ the content of rhombohedral phase attains ~60% at x(Co) + y(Mg) ~0.11, whereas for the $LaCo_xMg_{0.5x}Ga_{1-1.5x}O_{3-\delta}$ the quantity of rhombohedral phase at this concentration remains at the level of ~15%.

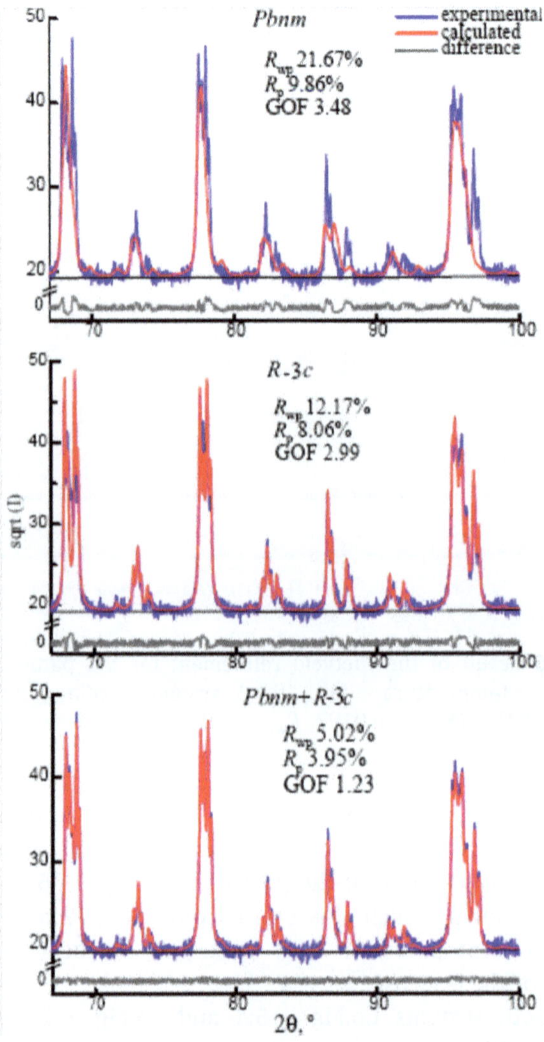

Figure 4.27 Calculation of pattern $LaGa_{1-1.2x}Co_xMg_{0.2x}O_3$, x(Co) = 0.0845, y(Mg) = 0.0197 by Rietveld method, using only *Pbnm*, only *R*-3*c*, and upon their joint presence (shows a range of angles 2θ = 67–100°).

The introduction of greater quantity of magnesium seems to stabilize the low symmetry orthorhombic phase owing to distortions of the structure. This is associated with the difference in the sizes of Mg^{2+} and Ga^{3+} atoms ($r(Mg^{2+})_{VI}$ = 0.74 Å vs. $r(Ga^{3+})_{VI}$ = 0.62 Å)[88] and the emergence of greater number of oxygen vacancies.

The changes in the unit cell volume for both phases are given in Fig. 4.28 (in inset). For the Co:Mg = 5:1 system the unit cell volume of orthorhombic and rhombohedral phases decreases nonlinearly as the quantity of doping elements increases. For the Co:Mg = 2:1 system, we observe almost linear decrease in the unit cell volume of orthorhombic phase, and nonlinear—for rhombohedral phase.

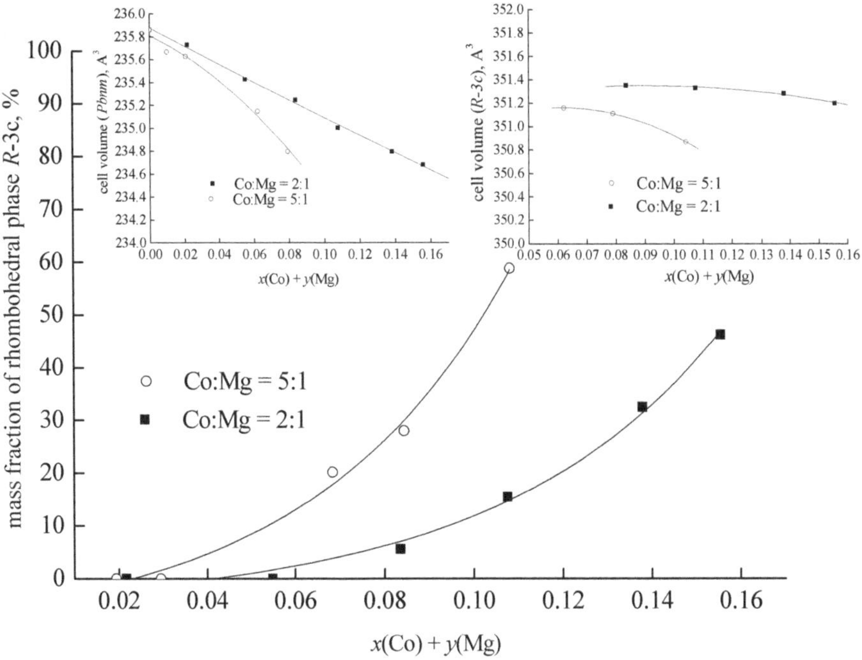

Figure 4.28 Mass fraction of rhombohedral phase (R-3c) vs. the doping element content for both systems. In inset: changes in the unit cell volume of orthorhombic and the rhombohedral phases.

The changes in the unit cell parameters for the Co:Mg = 5:1 and Co:Mg = 2:1 systems for orthorhombic phase are given in Fig. 4.29, and for rhombohedral phase—in Fig. 4.30. The trends in the changes of parameters are similar—in the orthorhombic phase a, b, and c monotonously and linearly decrease as the dopant concentration increases, and for rhombohedral phase a parameter decreases nonlinearly and c parameter increases nonlinearly. These data are in good agreement with the data of ref. 89, where the $LaGa_{1-x-y}Co_xMg_yO_{3-\delta}$ ($x = 0.6, y = 0.1; x = 0.4,$

$y = 0.2$) solid solutions were studied. The obtained values ($a = 5.469$ Å, $c = 13.175$ Å; $a = 5.492$ Å, $c = 13.249$ Å) are lower than in our work since the level of doping in ref. 89 is essentially higher.

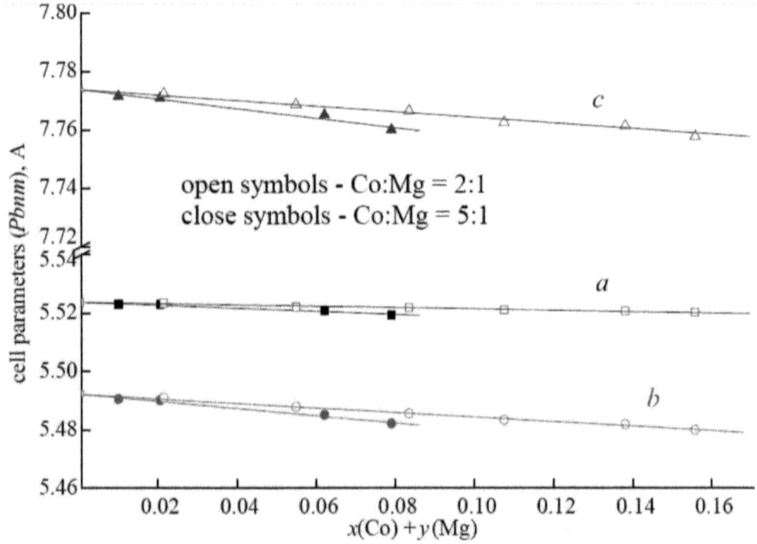

Figure 4.29 Plots of the unit cell parameters of orthorhombic phase vs. concentration for both systems.

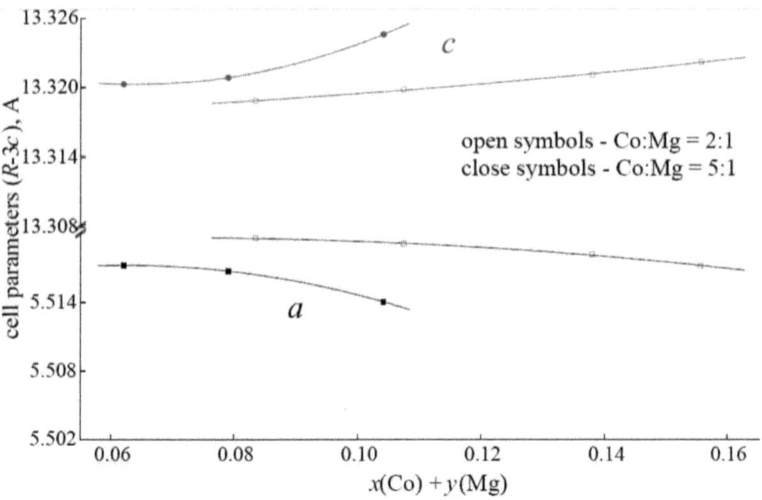

Figure 4.30 Plots of the unit cell parameters of rhombohedral phase vs. concentration for both systems.

4.5.3.2 Magnetic features

An interesting feature of these systems is the dependence of susceptibility on the magnetic field strength for the whole concentration interval. Such a phenomenon we observed upon magnetic dilution of lanthanum gallate containing chromium, magnesium and(or) strontium (see Section 4.4.1 and refs. 74, 75, and 83]). This dependence testifies for the presence of highly nuclear clusters of paramagnetic atoms in magnetically diluted systems rather than for long magnetic order.

As has been shown earlier for chromium-containing systems, the sizes of clusters attain 1–1.5 nm, i.e., they contain no less than 20 paramagnetic atoms. Moreover, the examination of magnetization suggests a super paramagnetic behavior of our systems.

From Fig. 4.31, where the dependence of molar magnetization on H/T is given, it is seen that the magnetizations at various temperatures and fields coincide. This is one of two requirements of superparamagnetism.

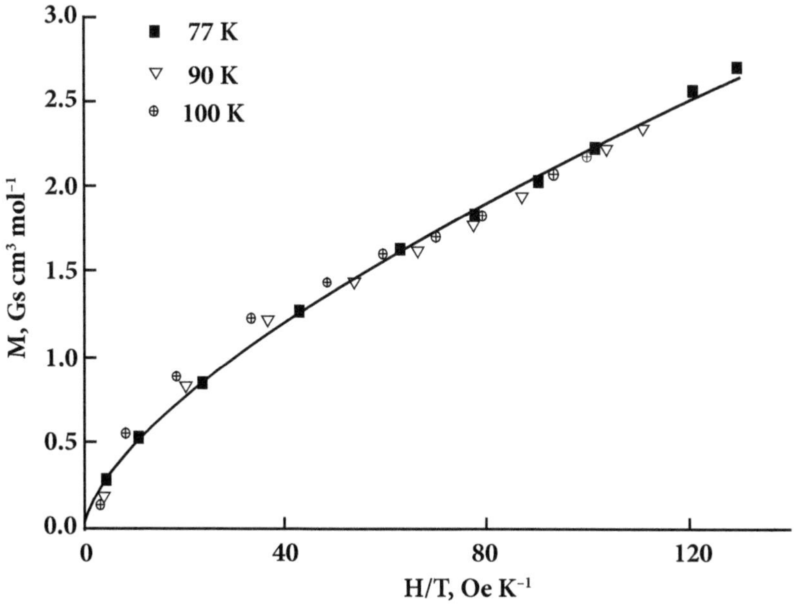

Figure 4.31 Plots of molar magnetization vs. H/T for $LaCo_xMg_{0.2x}Ga_{1-1.2x}O_3$ $x = 0.0165$.

Belov–Arrott plots testifies for the absence of spontaneous magnetization in our systems, since the thermodynamic coefficient α is positive everywhere (Fig. 4.32a,b).

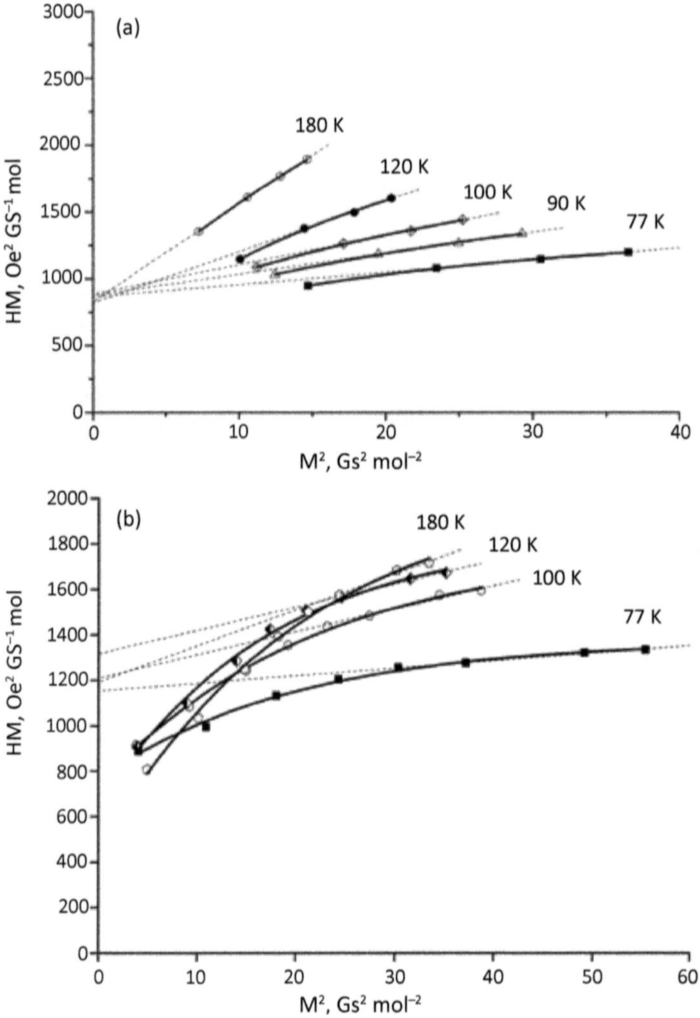

Figure 4.32 Belov–Arrott plots for (a) $LaCo_xMg_{0.2x}Ga_{1-1.2x}O_3$, x = 0.0165; (b) La $Co_xMg_{0.5x}Ga_{1-1.5x}O_3$, x = 0.0359.

Therefore, we may assume that in these systems as in chromium-containing systems there exist the clusters of high nuclearity.

To calculate the paramagnetic component of magnetic susceptibility we used the values of specific susceptibility extrapolated to infinitely high field strength ($1/H = 0$). The inverse paramagnetic component vs. temperature were plotted for the Co:Mg=5:1 and Co:Mg=2:1 systems, respectively (Fig. 4.33a,b).

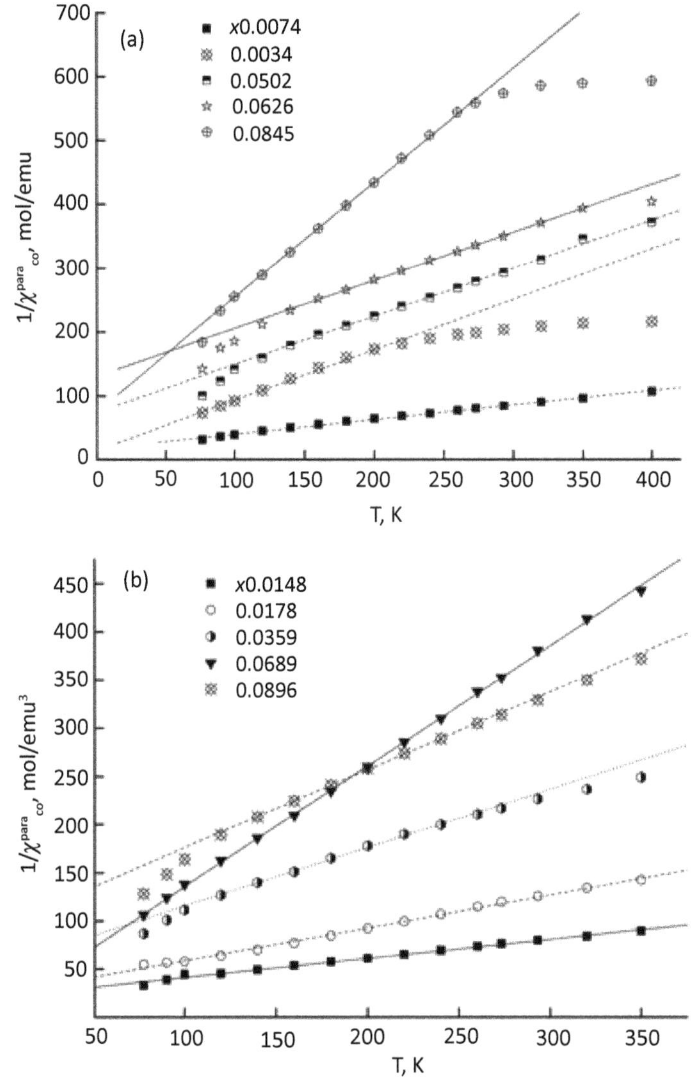

Figure 4.33 (a, b) Temperature dependence of inverse paramagnetic susceptibility for (a) $LaCo_xMg_{0.2x}Ga_{1-1.2x}O_3$ and (b) $LaCo_xMg_{0.5x}Ga_{1-1.5x}O_3$.

These dependencies demonstrate a deviation from Curie–Weiss law at low and at high temperatures. The deviations at high temperatures are typical for a contribution of temperature-independent paramagnetism, most probably Pauli paramagnetism of conductivity electrons, since it increases as the cobalt concentration increases in the solid solutions, i.e., as the concentration of e_g electrons in the forbidden band of lanthanum gallate increases.

The deviations of the dependencies from Curie–Weiss law at low temperatures seem to be determined by the character of exchange interactions between cobalt atoms and are typical for ferrimagnets. This is the first indication to a certain competition of ferro- and antiferromagnetic exchange interactions with those first prevailing.

Concentration dependencies of paramagnetic component and temperature dependencies of the effective magnetic moment are given in Figs. 4.34 and 4.35, respectively.[87]

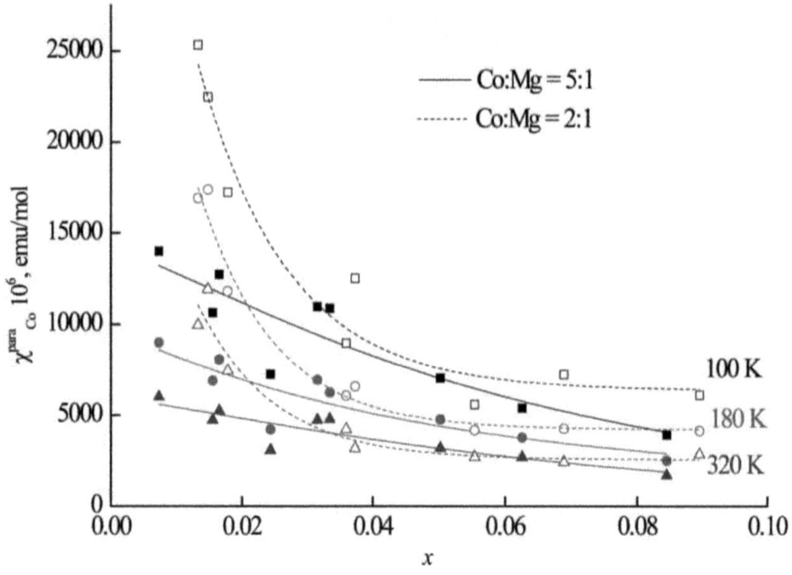

Figure 4.34 Concentration dependencies of paramagnetic component of magnetic susceptibility for $LaCo_xMg_{0.2x}Ga_{1-1.2x}O_3$ and $LaCo_xMg_{0.5x}Ga_{1-1.5x}O_3$ at three temperatures.

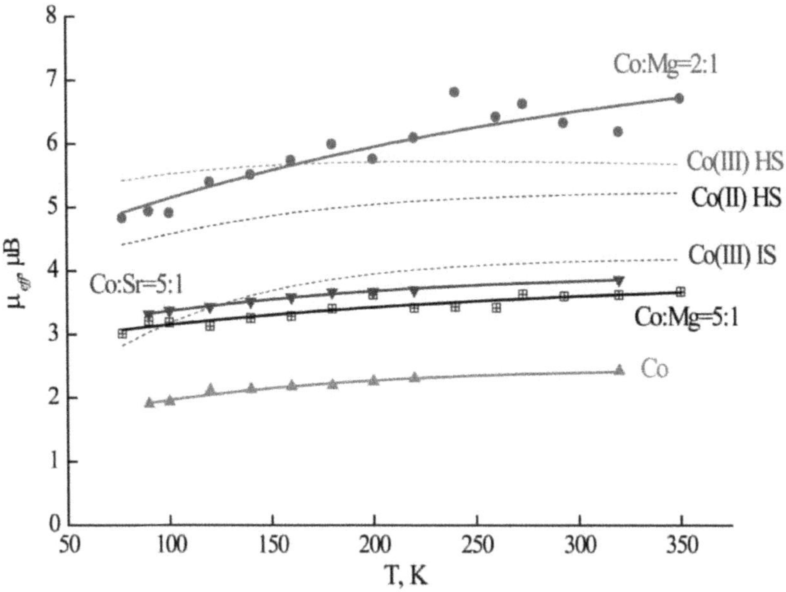

Figure 4.35 Temperature dependencies of the effective magnetic moment at the infinite dilution for various cobalt-containing lanthanum gallate systems.

The isotherms of paramagnetic susceptibility show that the exchange interactions in our solid solutions are antiferromagnetic. However, it must be emphasized that the isotherms lie essentially higher than for the samples containing no magnesium and the higher, the higher is magnesium content in the system.

Temperature dependencies of the effective magnetic moment extrapolated to the infinite dilution and their comparison with the $LaCo_xGa_{1-x}O_3$ and $La_{1-0.2x}Sr_{0.2x}Co_xGa_{1-x}O_3$ systems described earlier are the most interesting for conclusions about the states of cobalt atoms and exchange interactions between them.

In Fig. 4.35 the theoretical dependencies of magnetic moments for single cobalt atoms in the oxidation states 2 and 3 are given and also the dependencies for Co^{3+} in various spin states.

The matter is that for Co^{3+} in an octahedron of oxygen atoms three ground states can exist: $S = 0$, $^1A_{1g}$—low spin, $S = 2$, $^5T_{2g}$—high spin, and given substantial distortions in the nearest surrounding, which can be expected in our systems due to the vacancies and larger magnesium atoms in the same sites as cobalt, the state with the intermediate spin $S = 1$, $^3T_{1g}$.[65]

As is seen from Fig. 4.35, for the system with Co:Mg = 2:1, the effective magnetic moment cannot be described as belonging to single cobalt atoms, no matter in which valence state there were. Its values lie higher, especially at high temperatures.

For the systems with Co:Mg = 5:1, the effective magnetic moments lie lower and almost coincide with the values obtained for the systems containing strontium (Sections 4.1.3, 4.3.1.3, Fig. 4.1d). Those latter were described as belonging to single cobalt atoms partially in the low-spin state and partially in the state of spin equilibrium (low spin–high spin). However, we cannot describe the systems with magnesium using the same approach, since there is the dependence of susceptibility on the field strength.

The dependence of susceptibility on the field strength appears as the result of the solid solution synthesis, but is not the result of introducing some ferromagnetic admixtures. A similar dependence was observed for the $La_{1-0.2x}$ $Sr_{0.2x}Cr_x$ $Ga_{1-1.2x}Mg_{0.2x}O_{3-\delta}$ solid solutions (Section 4.4.1), therefore, it is an integral property of our systems.

This means that in our solid solutions with the ratio Co:Mg = 2:1, where it is evident, and even with the ratio Co:Mg = 5:1 at the infinite dilution sufficiently large clusters of cobalt atoms are preserved, the interaction between them being predominantly ferromagnetic with a certain contribution of antiferromagnetic exchange.

These clusters must contain also magnesium atoms and vacancies in the oxygen sublattice accompanying them.

For cobalt atoms such a situation may be associated with the fact that in the vicinity of magnesium atoms and vacancies inevitably appear the distortions in the cobalt surrounding resulting in its transfer to the state with the intermediate spin. The neighboring cobalt atoms appear to be high spin and the exchange between two such neighbors appear to be ferromagnetic.

4.6 Conductivity in the Systems with Magnesium and an Increased Content of Strontium

4.6.1 Chromium-Containing Systems

For all the systems under study given low concentrations of the doping elements, a break is observed in the plots $lg\sigma$ - T^{-1}. The activation energy changes, which testifies for a change in the mechanism of the charge transfer: At low temperatures the electronic component of conductivity prevails, at high temperatures the ionic conductivity switches on. An example for chromium-containing systems is given in Fig. 4.36 According to the data of thermal X-ray analysis this break is not associated with any polymorphic transformations and structural transitions.

An abrupt change in the mechanism of the charge transfer is observed in all cases for the most diluted solutions, since only in this concentration range the main role in conductivity is played by the highly nuclear clusters of chromium atoms including the oxygen vacancies in their composition. Coulomb field of clusters undoubtedly determine the whole energetics of the process of charge transfer for ionic conductivity (the typical energy for oxygen ion migration is 0.70 eV).

4.6.2 Nickel-Containing Systems

The conductivity of nickel-containing systems is similar, thus we show only the general patterns (Fig. 4.37).

In most cases as the concentration of the substituents increases, no changes in the activation energy occur, which seems to be associated with various disordering of vacancies in the matrix of lanthanum gallate.

On the basis of obtained dependencies we plotted the isotherms of conductivity (Fig. 4.38). Their run shows that the systems with the ratio M:Sr(Mg) = 5:1 have maximal conductivity. And what is more, from two systems M:Sr = 5:1 and M:Mg = 5:1, the conductivity of chromium-containing solid solution is greater.

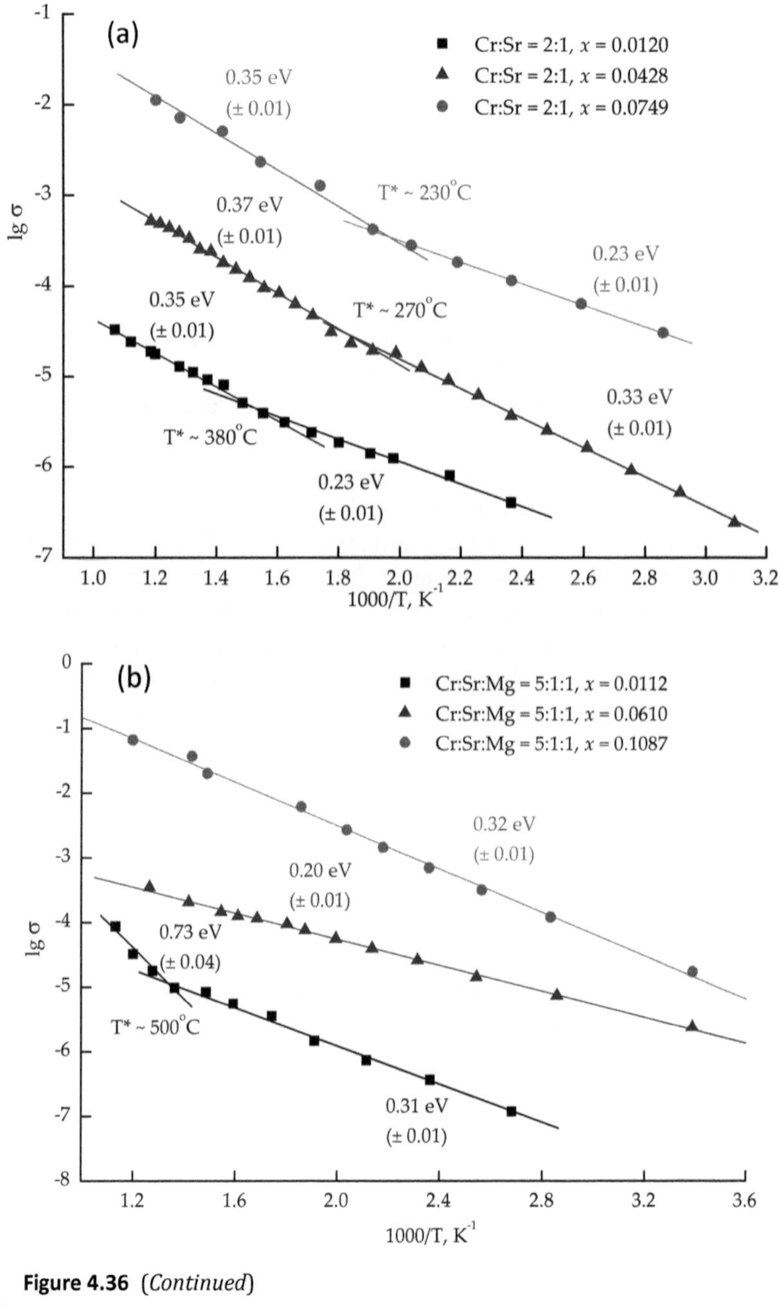

Figure 4.36 (*Continued*)

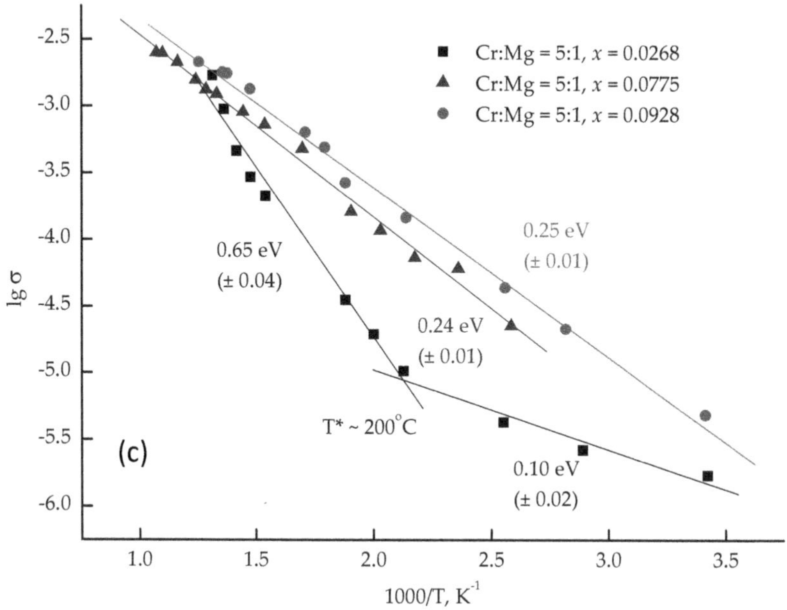

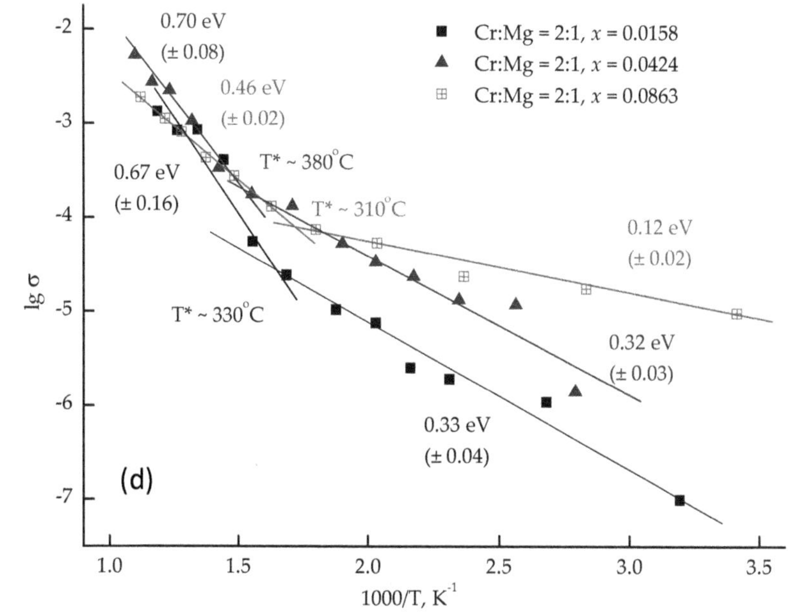

Figure 4.36 (*Continued*)

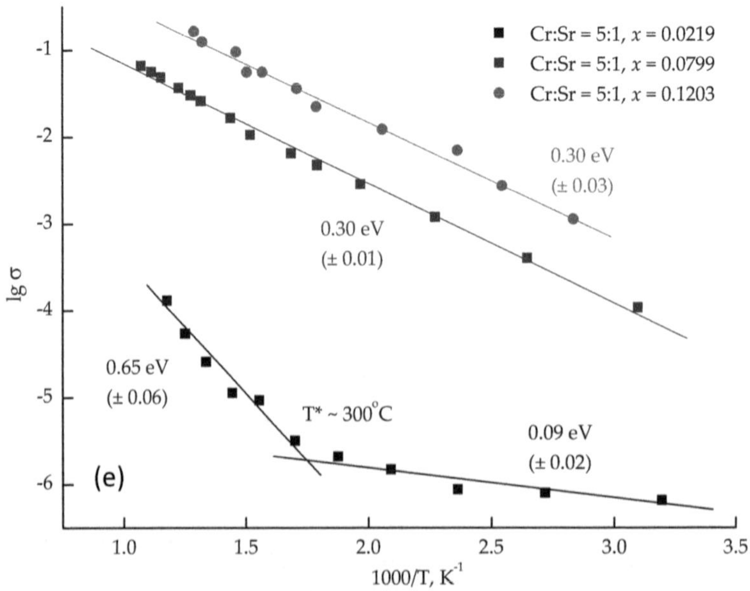

Figure 4.36 (a, b, c, d, e) Plots of logarithm of the total conductivity vs. inverse temperature for various chromium-containing systems and various concentrations of the solid solutions.

The reason seems to lie in the fact that magnesium is located in gallium sites and clustering in magnesium-containing systems is higher than for the systems-containing strontium. The conductivity of nickel-containing systems appears to be higher than that of corresponding chromium-containing systems, which is accounted for by the difference in the electron structure of transition elements- the fact discussed above.

4.7 Lanthanum Gallate Doped with Chromium and Calcium or Barium

We studied lanthanum gallate doped with chromium, but as a diamagnetic doping element were taken calcium or barium—$La_{1-0.2x}A_{0.2x}Cr_xGa_{1-x}O_{3-\delta}$ where A = Ca or Ba. The isotherms of paramagnetic component of magnetic susceptibility appear to lie higher than the isotherms for strontium-containing systems.

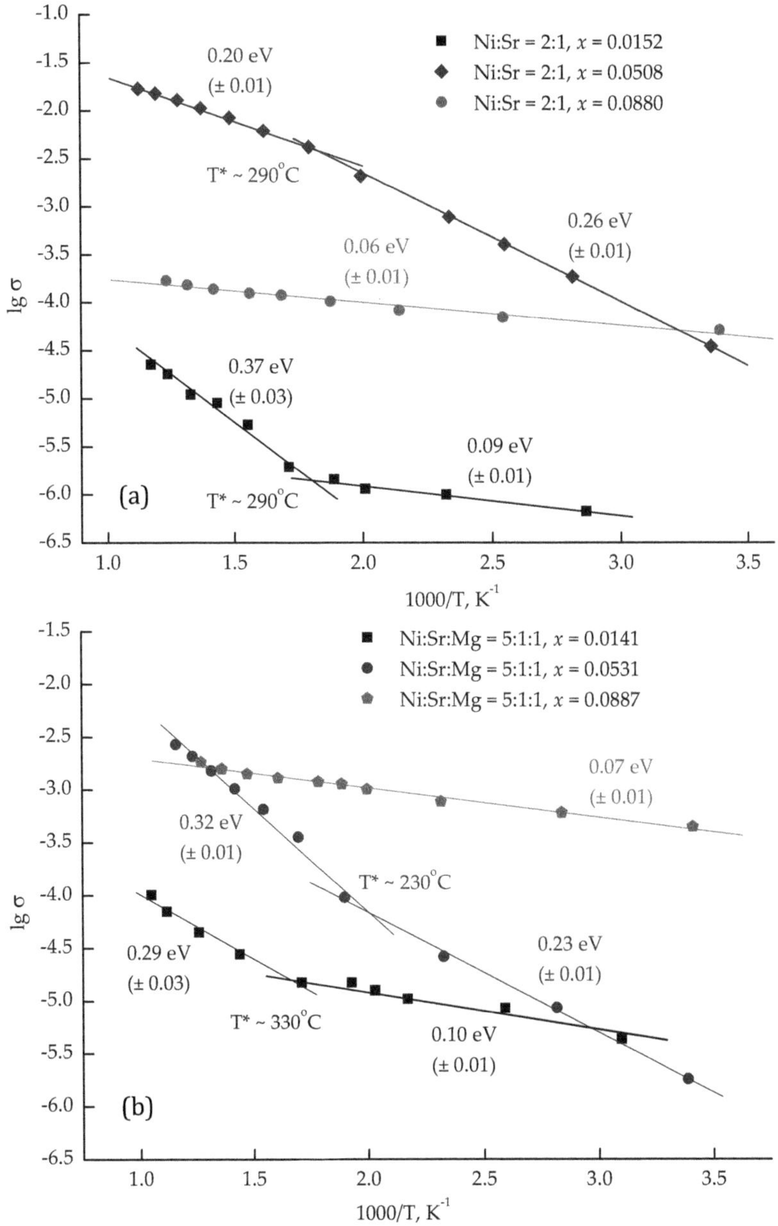

Figure 4.37 (*Continued*)

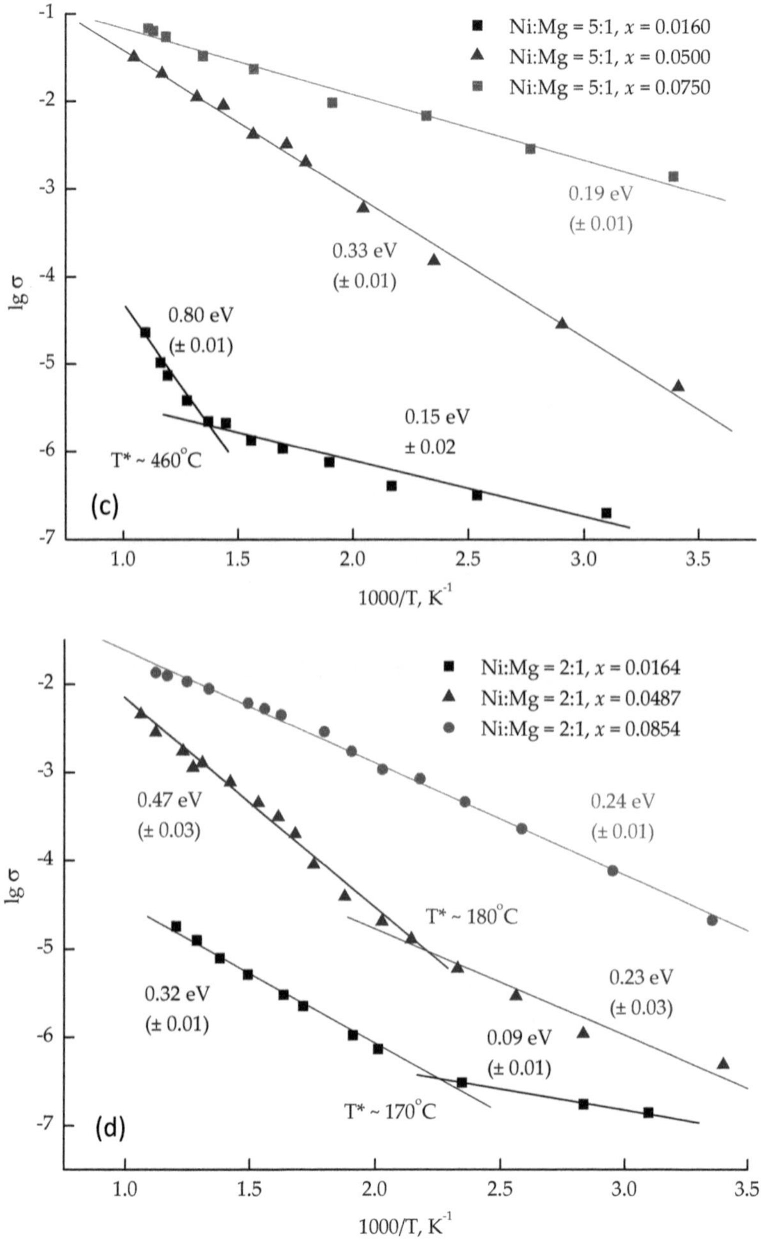

Figure 4.37 (*Continued*)

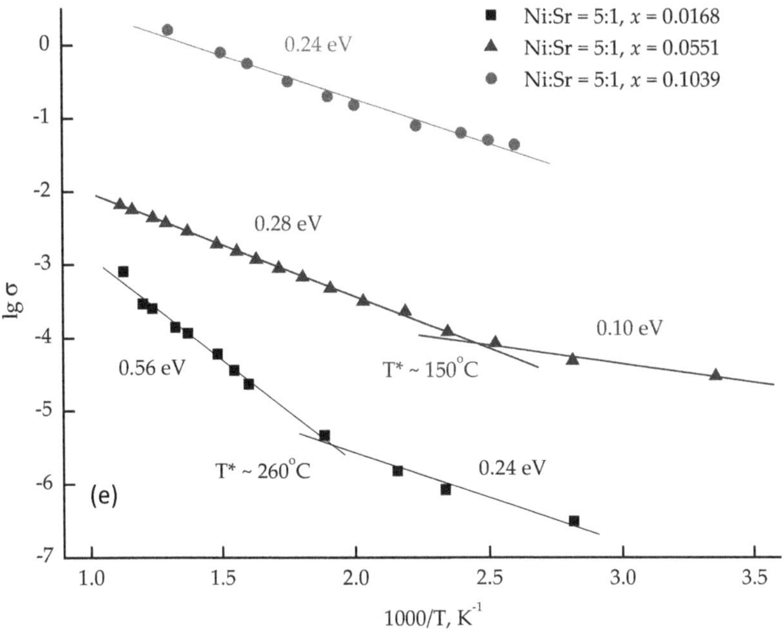

Figure 4.37 (a, b, c, d, e) Plots of logarithm of the total conductivity vs. inverse temperature for various nickel-containing systems and various concentrations of the solid solutions.

In the plots of inverse susceptibilities vs. temperature, the deviations from Curie–Weiss law are observed at low temperature pointing to the ferromagnetic behavior of these systems, just as we met for magnesium-containing systems and for the increased strontium content. The effective magnetic moments extrapolated to the infinite dilution are given in Fig. 4.39.

It is seen that we have the increased magnetic moments for both calcium and strontium-containing systems, which, together with the temperature dependencies of inverse susceptibility, points to the presence of clusters with competing ferro- and antiferromagnetic exchange, ferromagnetic exchange prevailing.

The reasons for such a nonmonotonous behavior of the solid solutions upon varying the diamagnetic heterovalent doping element seem to be the following. On passing along the series Ca–Sr–Ba the size of A atom in the solid solution formula

$La_{1-0.2x}A_{0.2x}Cr_xGa_{1-x}O_{3-\delta}$ increases, therefore the ionicity of A-O bond also increases, thus decreasing the ionicity of Cr-O bond. An increased covalence of the Cr-O bond results in an increased overlapping of *d*-orbitals with *p*-orbitals of oxygen.

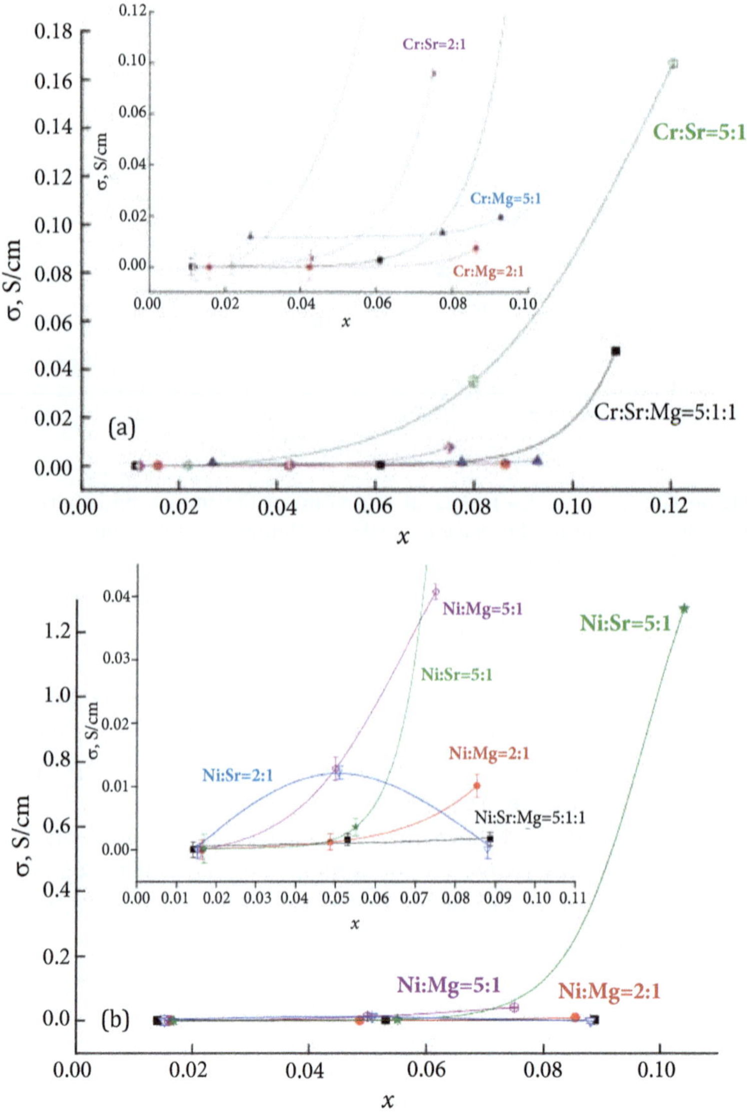

Figure 4.38 (a, b) Isotherms of conductivity for chromium and nickel systems with various concentration of strontium and (or) magnesium.

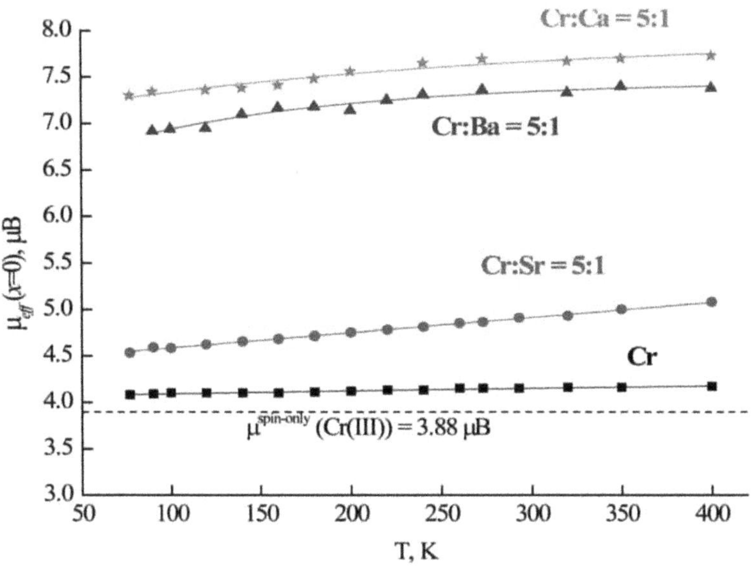

Figure 4.39 Temperature dependencies of the effective magnetic moment at the infinite dilution for $La_{1-0.2x}A_{0.2x}Cr_xGa_{1-x}O_{3-\delta}$ (A = Ca or Ba) systems.

This last circumstance must result in an increase in antiferromagnetic exchange and enhancement of clustering in barium-containing systems. In contrast to this in calcium-containing systems, where Cr-O bond has a more ionic character, clustering must be lower, but the exchange may substantially more ferromagnetic, which shows itself in greater effective magnetic moments. This means that we have two factors acting in opposite directions.

We may suggest with certainty that it is in the case of strontium an optimal relationship between the size of clusters and sufficiently strong ferromagnetic exchange is achieved, which is responsible for the stability of the system as a whole. This seems to be the reason for using strontium as a diamagnetic doping element in SOFC.

4.8 Conclusion

A detailed study of electron structure of a series of systems brings about the concept of the role of clustering both in the stability and in electrophysical performance of doped lanthanum gallate.

First, clustering in lanthanum gallate appears to be greater than in aluminate, which could be accounted for by the fact that aluminum being essentially smaller results in strong tensions in the lattice on introduction of larger elements.

Introduction of strontium rapidly results in the formation of K_2NiF_4 type structure—$LaSrAlO_4$. Small size of oxygen octahedra prevents clustering of 3*d*-elements $r(Al^{3+})_{VI}$ = 0.530 Å, $r(Cr^{3+})_{VI}$ = 0.615 Å, whereas $r(Ga^{3+})_{VI}$ = 0.620 Å.[87]

The study of magnetic characteristics and electrophysical performance of the lanthanum gallate doped with 3*d*-elements and heterovalent substituents (Sr, Mg) revealed the following. Introduction of greater than 5:1 quantity of strontium relative to the transition element and introduction of magnesium result in some changes in the electron structure of these solids—the clusters of high nuclearity are formed ($n > 20$) from transition metal atoms. The character of the exchange within clusters is determined by the electron structure of a transition element.

A correlation between the quantity of introduced doping element, its nature and magnetic characteristics of the systems accounts for the fact that the formation of clusters includes both sublattices—cationic and anionic, hence the clusters include magnetic atoms, diamagnetic heterovalent dopants and vacancies accompanying them.

The isotherms of conductivity also point to a correlation for all the systems under study: An increase in the concentration of heterovalent substituent results in a decrease in the conductivity, which can be associated with the vacancies being blocked within the high nuclear clusters (Fig. 4.38).

Therefore, the formation of a certain quantity of clusters of paramagnetic atoms is necessary for stabilization of the defect structure of lanthanum gallate. However, an increase in their sizes and fraction results in a decrease in the conductivity, which may be accounted for by the influence of Coulomb field of a cluster on the process of oxygen ion migration (Scheme 4.1).

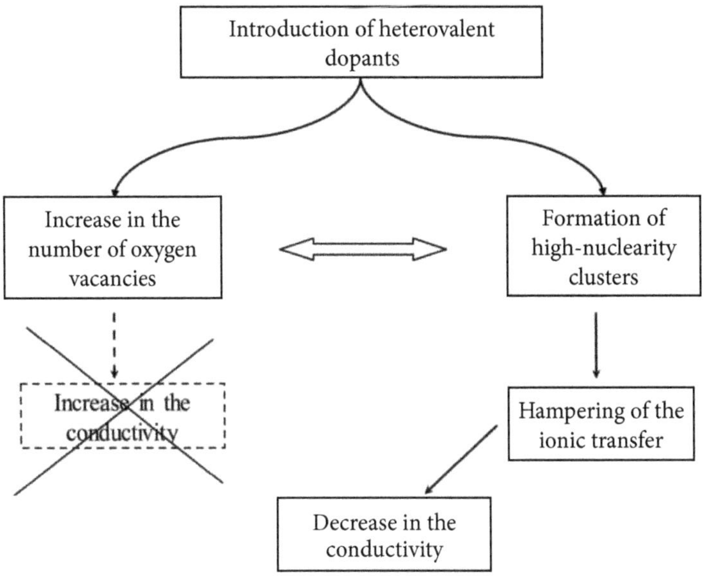

Scheme 4.1 Influence of heterovalent substitution on the conductivity of doped lanthanum gallate.

This study allows the theoretical basis to be found for the ratio M:Sr(Mg) = 5:1, which according to the published data is often used in practice.

References

1. Ishihara T, Matsuda H, Takita Y (1994) Doped $LaGaO_3$ perovskite type oxide as a new oxide ionic conductor, *J Am Chem Soc*, **116**, 3801–3803.
2. Kuroda K, Hashimoto I, Adachi K, Akikusa J, Tamou Y, Komada N, Ishihara T, Takita Y (2000) Characterization of solid oxide fuel cell using doped lanthanum gallate, *Solid State Ionics*, **132**, 199–208.
3. Du Y, Sammes NM (2001) Fabrication of tubular electrolytes for solid oxide fuel cells using strontium- and magnesium-doped $LaGaO_3$ materials, *J Eur Ceram Soc*, **21**, 727–735.
4. Enoki M, Yan J, Matsumoto H, Ishihara T (2006) Oxide ion conductivity in Fe and Mg doped $LaGaO_3$ and application for solid oxide fuel cells, *Solid State Ionics*, **177**, 2053–2057.

5. Ishihara T, Tabuchi J, Ishikawa S, Yan J, Enoki M, Matsumoto H (2006) Recent progress in $LaGaO_3$-based solid electrolyte for intermediate SOFCs, *Solid State Ionics*, **177**, 1949–1953.

6. Jacobson AJ (2010) Materials for solid oxide fuel cells, *Chem Mater*, **22**(3), 660–674.

7. Skinner SJ, Kilner JA (2003) Oxygen ion conductors, *Mater Today*, **6**(3), 30–37.

8. Kajitani M, Matsuda M, Hoshikawa A, Oikawa K, Torii S, Kamiyama T, Izumi F, Miyake M (2003) Neutron diffraction study on lanthanum gallate perovskite compound series, *Chem Mater*, **15**(18), 3468–3473.

9. Tas AC, Majewski PJ, Aldinger F (2000) Chemical preparation of pure and strontium- and/or magnesium-doped lanthanum gallate powders, *J Am Ceram Soc*, **83**(12), 2954–2960.

10. Datta P, Majewski P, Aldinger F (2007) Structural studies of Sr- and Mg-doped $LaGaO_3$, *J Alloys Compd*, **438**, 232–237.

11. Zheng F, Bordia RK, Pederson LR (2004) Phase constitution in Sr and Mg doped $LaGaO_3$ system, *Mater Res Bull*, **39**, 141–155.

12. Jasiolek G, Dabkowska H (1990) X-ray emission studies of some $ReGaO_3$ single-crystals, *J Less Common Met*, **160**, 79–84.

13. Wang Y, Liu X, Yao G-D, Liebermann RC, Dudley M (1991) High-temperature transmission electron microscopy and X-ray diffraction studies of twinning and the phase transition at 145°C in $LaGaO_3$, *Mater Sci Eng*, **A132**, 13–21.

14. Kobayashi J, Tazoh Y, Sasaura M, Miyazawa S (1991) Structural analysis of lanthanum gallate, *J Mater Res*, **6**(1), 97–100.

15. Marti W, Fischer P, Altorfer F, Scheel HJ, Tadin M (1994) Crystal structures and phase transitions of orthorhombic and rhombohedral $RGaO_3$ (R = La, Pr, Nd) investigated by neutron powder diffraction, *J Phys Condens Matter*, **6**(1), 127–135.

16. Slater PR, Irvine JTS, Ishihara T, Takita Y (1998) The structure of the oxide ion conductor $La_{0.9}Sr_{0.1}Ga_{0.8}Mg_{0.2}O_{2.85}$ by powder neutron diffraction, *Solid State Ionics*, **107**, 319–323.

17. Slater PR, Irvine JTS, Ishihara T, Takita Y (1998) Structure of lanthanum gallate electrolytes at high temperatures, 3rd European Solid Oxide Fuel Cell Forum, Nantes, France, pp. 387–396.

18. Slater PR, Irvine JTS, Ishihara T, Takita Y (1998) High-temperature powder neutron diffraction study of the oxide ion conductor $La_{0.9}Sr_{0.1}Ga_{0.8}Mg_{0.2}O_{2.85}$, *J Solid State Chem*, **139**(1), 135–143.

19. Majewski P, Rozumek M, Tas C, Aldinger F (2002) Processing of (La,Sr)(Ga,Mg)O_3 solid electrolyte, *J Electroceram*, **8**, 65–73.

20. Feng M, Goodenough JB (1994) A superior oxide-ion electrolyte, *Eur J Solid State Inorg Chem*, **31**, 663–672.

21. Lerch M, Boysen H, Hansen T (2001) High-temperature neutron scattering investigation of pure and doped lanthanum gallate, *J Phys Chem Solids*, **62**(3), 445–455.

22. Wang WL, Lu HY (2006) Twinning induced by the rhombohedral to orthorhombic phase transition in lanthanum gallate ($LaGaO_3$), *Phys Chem Miner*, **33**(7), 435–444.

23. Cheng J, Navrotsky A (2004) Energetics of magnesium, strontium and barium doped lanthanum gallate perovskites, *J Solid State Chem*, **177**(1), 126–133.

24. Khan MS, Islam MS, Bates DR (1998) Dopant substitution and ion migration in the $LaGaO_3$-based oxide ion conductor, *J Phys Chem B*, **102**(17), 3099–3104.

25. Islam MS (2002) Computer modeling of defects and transport in perovskite oxides, *Solid State Ionics*, **154–155**, 75–85.

26. Malavasi L, Fisher CAJ, Islam MS (2010) Oxide-ion and proton conducting electrolyte materials for clean energy applications: Structural and mechanistic features, *Chem Soc Rev*, **39**, 4370–4387.

27. Kharton VV, Viskup AP, Yaremchenko AA, Baker RT, Gharbage B, Mather GC, Figueiredo FM, Naumovich EN, Marques FMB (2000) Ionic conductivity of La(Sr)Ga(Mg, M)$O_{3-\delta}$ (M=Ti, Cr, Fe, Co, Ni): Effects of transition metal dopants, *Solid State Ionics*, **132**, 119–130.

28. Stevenson JW, Armstrong TR, Pederson LR, Li J, Lewinsohn CA, Baskaran S (1998) Effect of A-site cation nonstoichiometry on the properties of doped lanthanum gallate, *Solid State Ionics*, **113–115**, 571–583.

29. Polini R, Pamio A, Traversa E (2004) Effect of synthetic route on sintering behaviour, phase purity and conductivity of Sr- and Mg- doped $LaGaO_3$ perovskites, *J Eur Ceram Soc*, **24**, 1365–1370.

30. Li Z, Zhang H, Bergman B, Zou X (2006) Synthesis and characterization of $La_{0.85}Sr_{0.15}Ga_{0.85}Mg_{0.15}O_{3-\delta}$ electrolyte by steric entrapment synthesis method, *J Eur Ceram Soc*, **26**(12), 2357–2364.

31. Shi M, Xu Y, Liu A, Liu N, Wang C, Majewski P, Aldinger F (2009) Synthesis and characterization of Sr- and Mg-doped lanthanum gallate electrolyte materials prepared via the Pechini method, *Mater Chem Phys*, **114**, 43–46.

32. Datta P, Majewski P, Aldinger F (2007) Synthesis and microstructural characterization of Sr- and Mg-substituted $LaGaO_3$ solid electrolyte, *Mater Chem Phys,* **102**(2–3), 240–244.

33. Huang K, Goodenough JB (1998) Wet chemical synthesis of Sr- and Mg-doped $LaGaO_3$, a perovskite-type oxide-ion conductor, *J Solid State Chem,* **136**(2), 274–283.

34. Long NJ, Lecarpentier F, Tuller HL (1999) Performance of $La_{0.9}Sr_{0.1}Ga_{0.5}Ni_{0.5}O_3$ as a cathode for a lanthanum gallate fuel cell, *J Electroceram,* **3**, 399–407.

35. Trofimenko N, Ullmann H (1999) Transition metal doped lanthanum gallates, *Solid State Ionics,* **118**, 215–227.

36. Li S, Bergman B (2009) Doping effect on secondary phases, microstructure and electrical conductivities of $LaGaO_3$ based perovskites, *J Eur Ceram Soc,* **29**(6), 1139.

37. Maglia F, Anselmi-Tamburini U, Chiodelli G, Camurlu HE, Dapiaggi M, Munir ZA (2009) Electrical, structural, and microstructural characterization of nanometric $La_{0.9}Sr_{0.1}Ga_{0.8}Mg_{0.2}O_{3-\delta}$ (LSGM) prepared by high-pressure spark plasma sintering, *Solid State Ionics,* **180**, 36–40.

38. Huang K, Tichy RS, Goodenough JB (1998) Superior perovskite oxide-ion conductor: Strontium- and magnesium-doped $LaGaO_3$: I. Phase relationships and electrical properties, *J Am Ceram Soc,* **81**(10), 2565–2580.

39. Pathak S, Steinmetz D, Kuebler J, Payzant EA, Orlovskaya N (2009) Mechanical behaviour of $La_{0.8}Sr_{0.2}Ga_{0.8}Mg_{0.2}O_3$ perovskites, *Ceram Int,* **35**, 1235–1241.

40. Ha SB, Choa YH, Kang YC, Lee J-H (2010) Effect of oxide additives on the sintering behavior and electrical properties of strontium- and magnesium-doped lanthanum gallate, *J Eur Ceram Soc,* **30**, 2593.

41. Hsieh F-F, Okinaka N, Akiyama T (2009) Combustion synthesis of doped $LaGaO_3$ perovskite oxide with Fe, *J Alloys Compd,* **484**(1–2), 747–752.

42. Ullmann H, Trofimenko N (1999) Composition, structure and transport properties of perovskite-type oxides, *Solid State Ionics,* **119**(1–4), 1–8.

43. Baker RT, Gharbage B, Marques FMB (1998) Processing and electrical conductivity of pure, Fe- and Cr-substituted $La_{0.9}Sr_{0.1}GaO_3$, *J Eur Ceram Soc,* **18**, 105–112.

44. Politova ED, Stefanovich SYu, Aleksandrovskii VV, Kaleva GM, Mosunov AV, Avetisov AK, Sung JS, Choo KY, Kim TH (2005) The

lanthanum gallate-based mixed conducting perovskite ceramics, *Phys Stat Sol (c)*, **2**(1), 196–199.

45. Politova ED, Aleksandrovskii VV, Kaleva GM, Mosunov AV, Suvorkin SV, Zaitsev SV, Sung JS, Choo KY, Kim TH (2006) Mixed conducting perovskite-like ceramics on the base of lanthanum gallate, *Solid State Ionics*, **177**, 1779–1783.
46. Trofimenko N, Ullmann H (1999) Co-doped LSGM: Composition–structure–conductivity relations, *Solid State Ionics*, **124**, 263–270.
47. Ishihara T, Ishikawa S, Hosoi K, Nishiguchi H, Takita Y (2004) Oxide ionic and electronic conduction in Ni-doped $LaGaO_3$-based oxide, *Solid State Ionics* **175**, 319–322.
48. Colomer MT, Kilner JA (2011) Ni-doped lanthanum gallate perovskites: Synthesis and structural, microstructural, and electrical characterization, *Solid State Ionics*, **182**(1), 76–81.
49. Ishihara T, Akbay T, Furutani H, Takita Y (1998) Improved oxide ion conductivity of Co doped $La_{0.8}Sr_{0.2}Ga_{0.8}Mg_{0.2}O_3$ perovskite type oxide, *Solid State Ionics*, **113–115**, 585–591.
50. Ishihara T, Ishikawa S, Ando M, Nishiguchi H, Takita Y (2004) P_{O2} dependence of valence number of Co in $LaGaO_3$ and its influence on partial electronic and oxide ionic conductivity, *Solid State Ionics*, **173**(1–4), 9–15.
51. Glowacki M, Runka T, Domukhovski V, Diduszko R, Mirkowska M, Berkowski M, Dabrowski B (2011) Growth and characterization of perovskite $LaGaO_3$ crystals doped with Sr and Mn, *J Alloys Compd*, **509**, 1756–1759.
52. Ishihara T, Yamada T, Arikawa H, Nishiguchi H, Takita Y (2000) Mixed electronic-oxide ionic conductivity and oxygen permeating property of Fe-, Co- or Ni-doped $LaGaO_3$ perovskite oxide, *Solid State Ionics*, **135**(1–4), 631–636.
53. Yashima M, Nomura K, Kageyama H, Miyazaki Y, Chitose N, Adachi K (2003) Conduction path and disorder in the fast oxide-ion conductor $(La_{0.8}Sr_{0.2})(Ga_{0.8}Mg_{0.15}Co_{0.05})O_{2.8}$, *Chem Phys Lett*, **380**, 391–396.
54. Mori K, Onodera Y, Kiyanagi R, Richardson JW, Itoh K, Sugiyama M, Kamiyama T, Fukunaga T (2009) Structural and electrochemical properties of $La_{0.8}Sr_{0.2}GaFeO_3$, *Nucl Instr Met Phys Res A*, **600**, 328–331.
55. Yaremchenko AA, Shaula AL, Logvinovich DI, Kharton VV, Kovalevsky AV, Naumovich EN, Frade JR, Marques FMB (2003) Oxygen-ionic conductivity of perovskite-type $La_{1-x}Sr_xGa_{1-y}Mg_yM_{0.2}O_3$ (M = Fe, Co, Ni), *Mater Chem Phys*, **82**, 684–690.

56. Mabbs FE, Machin DJ (eds) (1973) *Magnetism and Transition Metal Complexes,* London: Chapman and Hall, New York.
57. Rakitin YuV, Kalinnikov VT (eds) (1994) *Modern Magnetochemistry,* Nauka, St. Petersburg [in Russian].
58. Fisher ME (1964) Magnetism in one-dimensional systems—the Heisenberg model for infinite spin, *Am J Phys,* **2**(5), 343–346.
59. Rakitin YuV, Starodub OR, Rakitina VM, Kalinnikov VT, Novotortsev VM (2005) Magnetic and thermodynamic properties of cyclic clusters $[s\text{–}S]_n$ with Alternating Spins: $s = 1/2$ and $S = 1$, $n \leq 8$ and $n \to \infty$, *Russ J Inorg Chem,* **50**(8), 1225–1231.
60. Smirnova NA (ed) (1973) *Methods of Statistical Thermodynamics in Physical Chemistry,* Moscow: Izd. Vysshaya Shkola [in Russian].
61. Kozheurov VA (1975) *Statistical Thermodynamics,* Moscow: Metallurgia [in Russian].
62. Chezhina NV (2006) Method of magnetic dilution in: *Physical Methods of Investigation of Inorganic Compounds,*ed., Nikol'skii, AB. Sankt-Petersburg: Izd. Academia [in Russian].
63. Zvereva I, Zueva F, Archaimbault M, Crespin J, Choisnet J, Lecomte L (1997) Crystallochemical, magnetic and electrical properties of the K_2NiF_4 type diluted solid solution $Y_{0.9}Ca_{1.1}Cr_yAl_{1-y}O_{4-\delta}$ ($y \leq 0.10$): Evidence for a partial $Cr^{3+} \to Cr^{4+}$ oxidation, *Mater Chem Phys,* **48**(2), 103–110.
64. Brach BYa, Bobrysheva NP, Reznitskii LA, Chezhina NV (1979) *Bull St Petersburg Univ,* **22**, 109.
65. Chezhina NV, Zharikova EV, Sidorova NYu (2010) Atom states and interatomic interactions in perovskite-like oxides: XXV. Magnetic dilution in the $La(Sr)MnO_3$-$LaGaO_3$ system, *Russ J Gen Chem,* **80**(2), 207–212.
66. Kalinnikov VT, Rakitin YuV (eds) (1980) *Introduction to Magneto-Chemistry. Method of Static Magnetic Susceptibility,* Nauka, Moscow [in Russian].
67. Martin RL, White AN (eds) (1968) in: *Transition Metal Chemistry* 4, Dekker, New York, p. 113.
68. Chezhina NV, Zharikova EV, Knyazev MN (2010) Atom states and interatomic interactions in complex perovskite-like oxides: XXVIII. Magnetic dilution in the $La(Sr)CoO_3$-$LaGaO_3$ system, *Russ J Gen Chem,* **80**(12), 2399–2404.
69. Kharton VV, Viskup AP, Naumovich EN, Lapchuk NM (1997) Mixed electronic and ionic conductivity of $LaCo(M)O_3$ (M=Ga, Cr, Fe or Ni):

I. Oxygen transport in perovskites $LaCoO_3$–$LaGaO_3$, *Solid State Ionics*, **104**(1–2), 67–78.

70. Chezhina NV, Zolotukhina NV, Bodritskaya EV (2005) State of atoms and interatomic interactions in complex perovskite-like oxides: XVIII. Magnetic dilution in the $LaCrO_3$-$LaGaO_3$ system, *Russ J Gen Chem*, **75**(8), 1167–1170.

71. Chezhina NV, Bodritskaya EV, Zhuk NA (2008) States of atoms and interatomic interactions in complex perovskite-like oxides: XXIII. Magnetic dilution in the La(Sr)NiO_3-$LaGaO_3$ system, *Russ J Gen Chem*, **78**(6), 1127–1132.

72. Chezhina NV, Bodritskaya EV, Zhuk NA, Bannikov VV, Shein IR, Ivanovskii AL (2008) Magnetic properties and electronic structure of the $LaGaO_3$ perovskite doped with nickel, *Phys Solid State*, **50**(11), 2121–2126.

73. Chezhina NV, Korolev DA, Sukharzhevskii SM, Glumov OV (2010) Atom states and interatomic interactions in perovskite-like oxides: XXVII. Influence of strontium concentration on interatomic interactions and conductivity in lanthanum gallates doped with strontium and chromium, *Russ J Gen Chem*, **80**(5), 915–919.

74. Korolev DA, Chezhina NV (2011) Atom state and interatomic interactions in complex perovskite-like oxides: XXIX. Influence of strontium concentration on special features of magnetic dilution in the La(Sr)CrO_3-$LaGaO_3$ system, *Russ J Gen Chem*, **81**(10), 2067–2073.

75. Korolev DA, Chezhina NV, Lytkina ZhA (2012) State of atoms and interatomic interactions in complex perovskite-like oxides: XXXI. Influence of magnesium concentration on chromium atoms state and interatomic interactions in lanthanum gallate doped with chromium and magnesium, *Russ J Gen Chem*, **84**(3), 354–359.

76. Ishikawa Y (1962) Magnetic properties of the ilmenite-hematite system at lower temperature, *J Phys Soc Japan*, **17**, 1835–1844.

77. Ishikawa Y (1962) Superparamagnetism in the $ZnFe_2O_4$-$NiFe_2O_4$ system, *J Phys Soc Japan*, **17**, 1877–1883.

78. Bajpai A, Banerjee A (2000) Superparamagnetism in polycrystalline $Li_{0.5}Ni_{0.5}O$ samples: Low-field susceptibility measurements, *Phys Rev B*, **62**, 8996–8998.

79. Belov KP (ed) (1959) *Magnetic Transformations* Izd. GIFML, Moscow [in Russian].

80. Bean CP, Livingston JD (1959) Superparamagnetism, *J Appl Phys*, **10**, 120S.

81. Kellerman DG, Gorshkov VS, Karelina VV (1999) Short-range order in cubic solid solutions $LiFe_{1-x}Ni_xO_2$ ($0 \leq x \leq 0.4$), *Russ J Inorg Chem*, **44**(7), 1078–1083.

82. Kellerman DG, Shalaeva EV, Gusev AI (2004) Cluster formation in $LiFe_{0.4}Ni_{0.6}O_2$, *Phys Solid State*, **46**(9), 1686–1692.

83. Chezhina NV, Korolev DA (2012) State of atoms and interatomic interactions in complex perovskite-like oxides: XXX. Influence of the nature of diamagnetic substituents on the dynamics of clustering in lanthanum gallate doped with strontium, chromium, and magnesium, *Russ J Gen Chem*, **82**(3), 347–353.

84. Naumovich EN, Kharton VV, Yaremchenko AA, Patrakeev MV, Kellerman DG, Logvinovich DI, Kozhevnikov VL (2006) Defect formation in $LaGa(Mg,Ni)O_{3-\delta}$: A statistical thermodynamic analysis validated by mixed conductivity and magnetic susceptibility measurements, *Phys Rev B*, **74**, 064105.

85. Yaremchenko AA, Kharton VV, Naumovich EN, Shestakov DI, Chukharev VF, Kovalevsky AV, Shaula AL, Patrakeev MV, Frade JR, Marques FMB (2006) Oxygen stoichiometry of $LaGa_{0.65}Mg_{0.15}Ni_{0.20}O_{3-\delta}$, *Solid State Ionics*, **177**, 549–558.

86. Howard CJ, Kennedy BJ (1999) The orthorhombic and rhombohedral phases of $LaGaO_3$—a neutron powder diffraction study, *J Phys Cond Matter*, **11**(6), 3229–3236.

87. Korolev DA, Chezhina NV, Lopatin SI (2015) Magnetic study of interatomic interactions, synthesis, structural and mass spectroscopy investigations of lanthanum gallate doped with cobalt and magnesium, *J Alloys Compd*, **624**, 53–59.

88. Shannon RD, Prewitt CT (1969) Effective ionic radii in oxides and fluorides, *Acta Cryst B*, **25**, 925–946.

89. Yaremchenko AA, Kharton VV, Viskup AP, Naumovich EN, Tikhonovich VN, Lapchuk NM (1999) Mixed electronic and ionic conductivity of $LaCo(M)O_3$ (M = Ga, Cr, Fe or Ni). V. Oxygen permeability of Mg-doped $La(Ga, Co)O_{3-\delta}$ perovskites, *Solid State Ionics*, **120**, 65–74.

Chapter 5

Magnetic Behavior of Multicomponent Bismuth Niobates and Bismuth Titanates, with Pyrochlore and Layered Perovskite-Type Structures

Irina V. Piir,[a] Mariya S. Koroleva,[a] Dmitry A. Korolev,[b] and Natalia V. Chezhina[b]

[a]*Institute of Chemistry Komi Science Center Ural Branch, Russian Academy Sciences, Pervomayskaya St. 48, Syktyvkar, 167982, Russia*
[b]*Department of General and Inorganic Chemistry, St. Petersburg State University, Universitetskaya nab. 7/9, St. Petersburg, 199034, Russia*

ipiir@mail.ru, n.chezhina@spbu.ru

5.1 Introduction

Multicomponent bismuth niobates and titanates related to the structural types of pyrochlore and layered perovskite attract attention in terms of both theoretical and experimental studies, in order to develop a scientific basis for obtaining multicomponent bismuth titanates with predictable properties and as new materials for electronic devices and devices for energy storage and information storage of the next generation.

Electronic Structure of Materials: Challenges and Developments
Edited by Natalia V. Chezhina and Dmitry A. Korolev

ISBN 978-981-4800-55-6 (Hardcover), 978-0-429-24287-8 (eBook)
www.panstanford.com

Materials based on new compounds with pyrochlore structure type show a variety of areas of their practical use, including the ability of individual compounds to immobilize actinides,[1] catalysis,[2–4] ferromagnetism,[5] dielectric properties,[6–11] and mixed ion conductivity.[12] Bismuth titanates with layered perovskite structure belonging to the family of the phases of Aurivillius retain ferroelectric properties over a wide range of temperatures and are promising materials in radio-, acousto- and optoelectronics, can be used to manufacture electronic capacitors, piezoelectric transducers, filters, hydroacoustic devices, and pyroelectric detectors of infrared radiation. Doped compositions on the basis of $Bi_4Ti_3O_{12}$ can be promising as lead-free ferroelectric and piezoelectric materials. For compounds containing transition elements, it is important to study their magnetic behavior in connection with the prospect of using them as multiferroics and materials for spintronics. The search for compounds that demonstrate interrelated magnetic and electrical effects continues to be an urgent problem in materials science and is currently activated due to real prospects emerging for the use of multiferroics in new areas of instrumentation and information storage devices. Such cooperative phenomena as ferroelectricity, ferro(antiferro)-magnetism, and possible interrelation and interaction between them in multifunctional materials result both from the local distribution of paramagnetic atoms in crystal structures and from the processes of atom disordering in complex structures such as, in particular, the pyrochlore structure. Different types of disordering are observed in the systems based on bismuth and niobium oxides with additional small size element (zinc, magnesium, 3*d*-elements) and in doped bismuth titanate pyrochlores. Bismuth niobates containing the above-mentioned elements and related to the structural type of pyrochlore differ from one another according to the stoichiometry, but all of them are characterized by the distribution of substituting cations over several (two) sites and by a displacement of bismuth atoms and oxygen atoms (O′) from the center of their sites in the structure of the ideal pyrochlore, $A_2B_2O_6O'$. The cation distribution in bismuth sites affects the number of vacancies in the bismuth sublattice and thus the stoichiometry of the compound. So, it becomes relevant to study the influence of the nature of the doping element on the cation distribution, on the local surroundings of atoms in substituted bismuth niobates and doped bismuth titanates, on the thermal stability of compounds with pyrochlore-type structure, and on their functional (electrical, magnetic and optical) properties. The possibilities of practical application of multicomponent bismuth titanates and niobates are determined by their structure, composition, stoichiometry, availability of vacancies in various sublattices, and local distribution of cations. It is necessary to study these systems systematically over a wide range of compositions and with the help of physical methods that give comprehensive information about the electronic structure of the compounds. A comparison between the characteristics of bismuth titanates and niobates with pyrochlore structure doped with the same elements (with possible substitution for two cation sites) and with layered perovskite structure (substitution for only one cation site) will reveal the influence of disordering on the properties of compounds with pyrochlore-type structure.

5.2 Magnetic Behavior of the Doped Titanates and Substituted Niobates of Bismuth with Layered Perovskite Structure

5.2.1 Structure of Layered Doped Bismuth Titanates and Bismuth Niobates

In the Bi_2O_3–TiO_2 system a $Bi_4Ti_3O_{12}$ mixed oxide with layered perovskite structure related to the phases of Aurivillius is formed when $n(Bi)/n(Ti) > 1$.[13] The family of Aurivillius phases includes bismuth titanates with the general formula $(Bi_2O_2)^{2+}(A_{m-1}B_mO_{3m+1})^{2-}$, where m is the number of perovskite-like layers between $(Bi_2O_2)^{2+}$ layers, A and B metal atoms in the I, II, III, and IV, V, VI oxidation states, respectively. Atoms A and B can occupy both types of the sites, and many Aurivillius phases elemental compositions can be expected. Possible compositions of Aurivillius phases are given in Table 5.1. The crystal structure of all Aurivillius phases consists of an alternation of perovskite-like layers $(A_{m-1}B_mO_{3m+1})^{2-}$ and bismuth-oxygen layers $Bi_2O_2{}^{2+}$. A-ions with large radii and rather low degrees of oxidation (Bi, Pb, Ba, Sr, Ca, K, Na, lanthanides) are located in the dodecahedral surrounding of oxygen atoms, and *B*-ions with small radii are in the highest oxidation degree and octahedral oxygen coordination. The value of *m* may vary in the range 1–5.

Table 5.1 Aurivillius phases based on bismuth titanate and niobate

Composition	Substitution element
$Bi_4Ti_3O_{12}$	Ti: Fe, Cr, Mn, Ni, Co
$Bi_5Nb_3O_{15}$	Bi: Na, lanthanides; Nb: Ta
$Bi_2SrNb_2O_9$	Sr: Ca, Ba, Pb; Nb: Ta

The stability of Aurivillius phases is associated with a high electronegativity of substituting atoms, which corresponds to their effective charge. The statistical distribution of 3*d*-elements (iron, chromium, manganese) in the perovskite block results in a weakening of the bonds between fluorite-like and perovskite-like layers. With increasing number of layers, starting from $m > 5$, there is a decrease in the thermal stability of the substituted compounds and a decrease in melting temperatures.[14–17] Thus, the decomposition of $Bi_{m+1}Fe_{m-3}Ti_3O_{3m+3}$ occurs with the following formation of phases with a lower number of layers in perovskite block belonging to the same homological series.[16] Bismuth titanate, $Bi_4Ti_3O_{12}$, belongs to Aurivillius phases with the number of octahedral layers in the perovskite-like block $m = 3$ (Fig. 5.1).

The structure of bismuth titanate, $Bi_4Ti_3O_{12}$, belongs to four spatial groups: *I*4/*mmm*, *B*2*cb*, *B*1*a*1, and *Fmmm*.[18–24] Studies of $Bi_4Ti_3O_{12}$ were carried out by neutron and X-ray diffraction and showed that the compound belongs to a orthorhombic system.[24,25] Bismuth titanate $Bi_4Ti_3O_{12}$ with tetragonal structure at $T < 600°C$ and an orthorhombic structure at 750°C is discussed in refs. 26 and 27. The change in the symmetry of the structure is due to the "ferroelectric-paraelectric" phase transformation, which occurs in $Bi_4Ti_3O_{12}$ at $T_c = 675°C$.[28] The space group for $Bi_4Ti_3O_{12}$ at $T < T_c$ is defined as tetragonal *I*4/*mmm* and as orthorhombic *Fmmm* at $T > T_c$.[27] The parameters of the unit cell for these $Bi_4Ti_3O_{12}$ structures are given in Table 5.2.

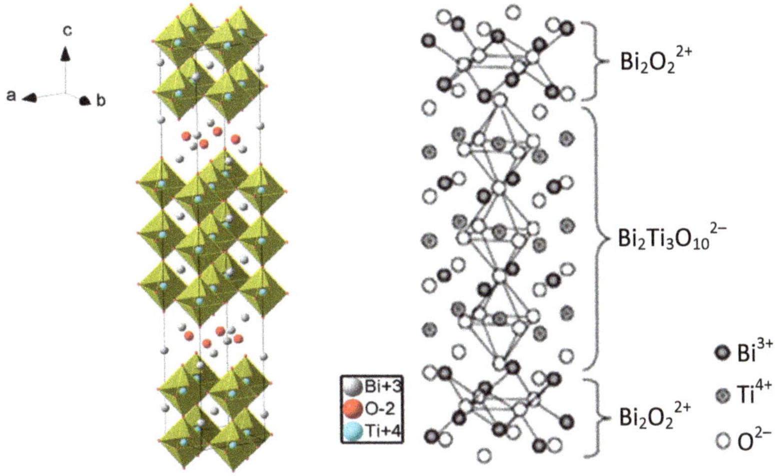

Figure 5.1 Structure of $Bi_4Ti_3O_{12}$.

Table 5.2 Space groups for $Bi_4Ti_3O_{12}$ and unit cell parameters (Z = 4, $\alpha = \beta = \gamma = 90°$)

	Unit cell parameters (Å)		
Space group	***a***	***b***	***c***
B2cb[24]	5.4432(5)	5.4099(5)	32.821(2)
B2cb[32]	5.44638(10)	5.40961(10)	32.8243(5)
Fmmm[29]	5.40256(15)	5.44065(15)	32.7929(9)
B1a1[31]	5.4475(8)	5.4082(9)	32.8066(6)
B1a1[33] (β = 89.95°)	5.445	5.412	32.817
I4/mmm[30]	3.8663(6)		33.648(8)

Bismuth niobate $Bi_5Nb_3O_{15}$ is characterized by an ordered alternation of bismuth-oxygen layers $(Bi_2O_2)^{2+}$ and perovskite-like blocks with a thickness of one and two niobium-oxygen octahedra so that the structure can be represented in the form $(Bi_2O_2)(NbO_4)(Bi_2O_2)(BiNb_2O_7)$.[34,35] The structure of the compound obtained as a result of the neutronographic study is shown in Fig. 5.2,[36] and the unit cell parameters for various space groups are given in Table 5.3.

Table 5.3 Space groups for $Bi_5Nb_3O_{15}$ and unit cell parameters ($Z = 2$)

	Unit cell parameters (Å)		
Space group	*a*	*b*	*c*
*Pnc*2[35]	5.464	5.390	21.01
*Pnc*2[36]	21.011(4)	5.473(1)	5.463(1)

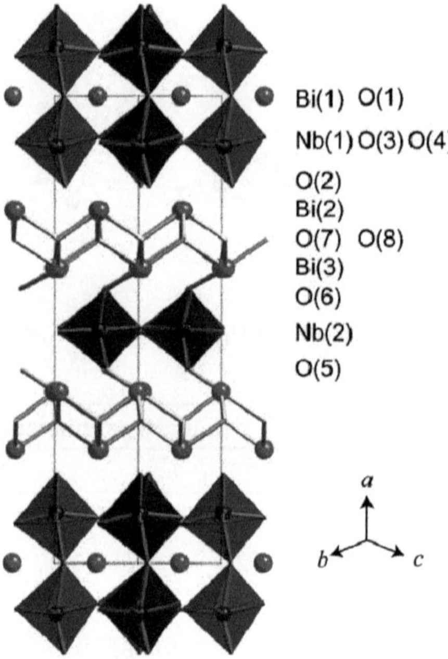

Figure 5.2 Structure of $Bi_5Nb_3O_{15}$.

5.2.2 Magnetic Properties of Doped Bismuth Niobates and Titanates

The magnetic behavior data for mixed niobates and bismuth titanates with layered perovskite structure are scarce and incomplete. The greatest part of the research is carried out for the $Bi_4Ti_3O_{12}$–$BiFeO_3$ system ($Bi_{m+1}Fe_{m-3}Ti_3O_{3m+3}$ compositions), where m is the number of layers in the perovskite block. Bismuth titanate $Bi_4Ti_3O_{12}$ is a diamagnetic compound. The magnetic properties of the $Bi_{m+1}Fe_{m-3}Ti_3O_{3m+3}$ compounds in $Bi_4Ti_3O_{12}$-$BiFeO_3$ system were studied in connection with the fact that bismuth ferrite $BiFeO_3$ demonstrates an electric and magnetic ordering, i.e., $BiFeO_3$ is a multiferroic with T_N = 370°C and is used to produce magnetoelectric materials. $Bi_{m+1}Fe_{m-3}Ti_3O_{3m+3}$ were found to be antiferromagnetic. Neel temperature (T_N) increases as the number of layers in the perovskite block increases, but its values remain below room temperature.[37] According to the results presented in ref. 38, $Bi_5Ti_3FeO_{15}$ is an antiferromagnetic with Neel temperature 80 K. However, the presence of a magnetic hysteresis loop for $Bi_5Ti_3FeO_{15}$ at room temperature was found, and magnetization linearly depends on the magnetic field strength.[39,40] The study of $Bi_5Ti_3FeO_{15}$ films also revealed a wide hysteresis loop with saturation magnetization $M_s \sim 0.21$ emu·g^{-1}.[40] The authors of the paper believe that hysteresis is ensured by the formation of ferromagnetic clusters due to the exchange interaction of Fe^{3+}-O-Fe^{3+}, directed perpendicularly to the $Bi_2O_2^{2+}$ layers. Magnetic properties also have been studied for chromium-containing bismuth titanates, $Bi_5Ti_3CrO_{15}$ with four layers in the perovskite-like block. It is a paramagnetic with antiferromagnetic behavior at $T < 150$ K.[41] The facts that the effective magnetic moment of chromium is less than the spin-only value [$\mu_{so}(Cr^{3+}) = 3.87$ μB] and Weiss constant ϑ is −25 K point to the antiferromagnetic exchange interaction between chromium atoms due to overlapping of chromium *d*-orbitals via *p*-orbitals of oxygen atoms at an angle of 180°. The magnetic hysteresis loop for $Bi_7Ti_3Fe_3O_{21}$ was detected at room temperature. Superexchange between Fe^{3+} at an angle of 180° must be antiferromagnetic, the ferromagnetic component can be observed only if the angle of overlapping of the iron *d*-orbitals via *p*-orbitals of oxygen is significantly distorted.[42,43] In a perovskite-like block paramagnetic atoms replace titanium atoms in the center of octahedra connected by apexes, and thus provide a superexchange. Therefore, the $Bi_4Ti_{3-x}Cr(Fe,Mn)_xO_{12-\delta}$ systems are a convenient model object for investigating exchange interactions in comparison with more complex systems of the pyrochlore structure based on titanium and bismuth niobates.

In the vast majority of experiments, weak ferromagnetic interactions in iron-containing bismuth titanate with layered perovskite structure are registered in film samples, practically nanostructured objects. The magnetic behavior of such systems seems to be determined to a greater extent by the short order interatomic interactions. This determines the prospects of using the magnetic dilution method in the study of substituted bismuth titanates $Bi_4Ti_{3-x}M_xO_{12}$ to obtain information about the behavior and interactions of paramagnetic atoms in the diamagnetic matrix over a wide concentration range at different dilution stages up to the hypothetical state of the infinite dilution ($x \rightarrow 0$).[44]

Magnetic dilution method (MDM) (see Chapter 1) was used to study the state of paramagnetic atoms in compounds. The fractions of single paramagnetic atoms and clusters—dimers in magnetically diluted chromium, iron, manganese-containing bismuth titanates with layered perovskite and pyrochlore structures were calculated on the basis of experimental susceptibilities measured over the whole concentration and temperature ranges. To perform these calculations we used the method of magnetic dilution,[44] the model of exchange interactions of the Heisenberg–Dirac–Van Vleck,[43] the model of exchange channels Rakitin and Eremin[45] (see Chapter 1).

For the Fe(III)-O-Fe(III) and Mn(II)-O-Mn(II) dimers, that is, for the pair $S_1 = S_2 = 5/2$, $g_1 = g_2 = 2$, $x = J_{dim}/kT$, the formula (Eq. 1.15, Table 5.3, Chapter 1) was used.[42]

The values of the exchange parameter J and the fraction of dimer clusters (a_{dim}) were found from the condition of the best agreement between the experimental and theoretical values of magnetic susceptibility while minimizing the function $\sum_i \sum_j (\chi_{ij}^{exp} - \chi_{ij}^{calc})^2$, where $\sum_i$ is the summation at all concentrations; $\sum_j$ the summation at all temperatures; and χ^{calc} and χ^{exp} the calculated and experimental values, respectively, of the paramagnetic component of magnetic susceptibility (Eq. 1.21, Chapter 1).

5.2.3 Magnetic Behavior of Chromium Containing Bismuth Titanates

The $Bi_4Ti_{3-x}Cr_xO_{12-\delta}$ solid solutions (x = 0.02–0.98) were obtained by solid-state reaction method. The amounts of bismuth, titanium, and chrome oxides have been calculated according to the stoichiometry of the samples. The starting oxides were mixed, thoroughly ground, pressed into tablets, and sintered at 600, 900, 1000, and 1100°C, consequently.[45]

The range of chromium concentrations (x), in which the $Bi_4Ti_{3-x}Cr_xO_{12-\delta}$ solid solutions are formed is $x < 0.98$ (Fig. 5.3). The unit cell parameters for the solid solution with $x = 0.4$ are $a = 5.413$ Å, $b = 5.443$ Å, $c = 32.847$ Å (S.G. *B*2*cb*). An increase in the chromium content in $Bi_4Ti_{3-x}Cr_xO_{12-\delta}$ to $x = 0.98$ results in the change in the space group to *Fmm*2: $a = 5.461$ Å, $b = 5.445$ Å, $c = 40.647$ Å.

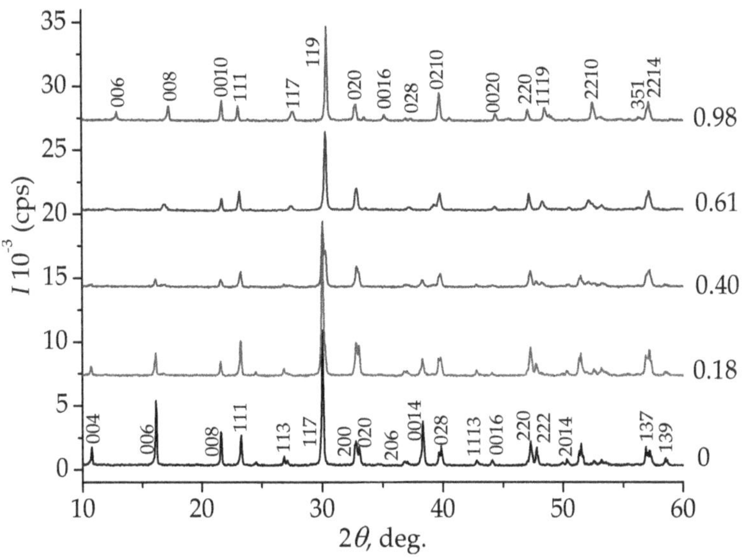

Figure 5.3 X-ray patterns for the $Bi_4Ti_{3-x}Cr_xO_{12-\delta}$ solid solutions (S.G. *B*2*cb*).

The magnetic properties of chromium-containing bismuth titanates with the structure of layered perovskite were studied for the $Bi_4Ti_{3-x}Cr_xO_{12-\delta}$ samples ($x = 0.02$–0.20).

The paramagnetic components of magnetic susceptibility and the effective magnetic moments at different temperatures and concentrations of chromium in the $Bi_4Ti_{3-x}Cr_xO_{12-\delta}$ solid solutions were calculated. For the $Bi_4Ti_{3-x}Cr_xO_{12-\delta}$ solid solutions, the dependence of the inverse paramagnetic component of magnetic susceptibility on temperature $\chi^{-1}(T)$ corresponds to Curie–Weiss law. The Weiss constant is negative (-65 K $\leq \theta \leq -32$ K) and points to antiferromagnetic exchange interactions in the structure of layered perovskite.

The effective magnetic moments for the diluted samples ($x \leq 0.03$) and for μ_{eff} ($x \to 0$) obtained upon extrapolating the concentration dependences of magnetic susceptibility to the infinite dilution are close to the spin-only values of 3.87 μB at the temperatures higher than 250 K. As the concentration of chromium in the solid solutions increases the magnetic moment decreases substantially. The dependence of the effective magnetic moment of chromium on the temperature at infinite dilution of chromium-containing bismuth titanate solid solutions is shown in Fig. 5.4.

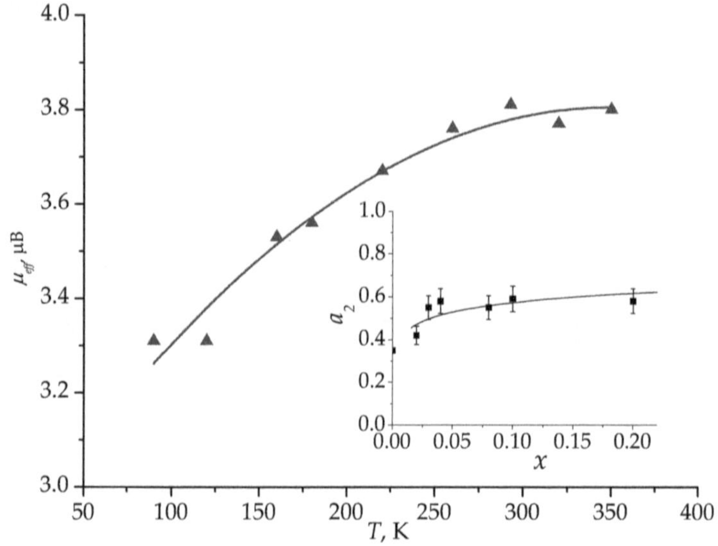

Figure 5.4 Temperature dependence of the effective magnetic moment of chromium in $Bi_4Ti_{3-x}Cr_xO_{12-\delta}$ at $x \to 0$. In the insert: fractions of dimers (a_2) in $Bi_4Ti_{3-x}Cr_xO_{12-\delta}$ ($J = -25$ cm^{-1}).

It must be noted that the effective magnetic moment (μ_{eff}) of $Bi_4Ti_{3-x}Cr_xO_{12-\delta}$ at $x \to 0$ depends on temperature, whereas such a dependence is impossible for single Cr^{3+} atoms ($^4A_{2g}$).[42] Consequently, it remains to be assumed that even at the infinite dilution no complete disaggregation of chromium atoms occurs, certain aggregates are preserved, perhaps Cr^{3+}-O-Cr^{3+} antiferromagnetically bonded (since μ_{eff} increases as the temperature increases). The calculation of cluster fractions (dimers) was carried out in the context of the above-mentioned methods (see Eq. 1.21, Chapter 1). Assuming the presence of both: single chromium atoms and Cr^{3+}-O-Cr^{3+} dimers in diluted solid solutions, the fractions of single chromium atoms (monomers) and dimers in the diluted solid solutions were determined. Corresponding equations were used in the calculations of dimer fractions. For the magnetic moment of the monomer spin-only effective magnetic moment Cr^{3+} (3.87 μB) was taken. A corresponding equation for the pair of chromium atoms $S_1 = S_2 = 3/2$, $g_1 = g_2 = 2$ and $x = J/kT$ was used to calculate the magnetic moment of the dimers.

The results of determining the fractions of dimers with an optimal exchange parameter ($J = -25$ cm^{-1}) and the fraction of single chromium atoms for solid solutions and for the hypothetical state of infinitely diluted solutions ($x \to 0$) are given in the insert in Fig. 5.4.[45] Experimental and theoretical values of the paramagnetic component of magnetic susceptibility are given in Fig. 5.5. We can assume the presence of a fairly high fraction of chromium dimers (a_2) even in the infinitely diluted solution of bismuth titanate–layered perovskite ($a_2 \approx 0.35$) and the existence of sufficiently strong interactions between chromium atoms, which can no longer be caused only by magnetic exchange.

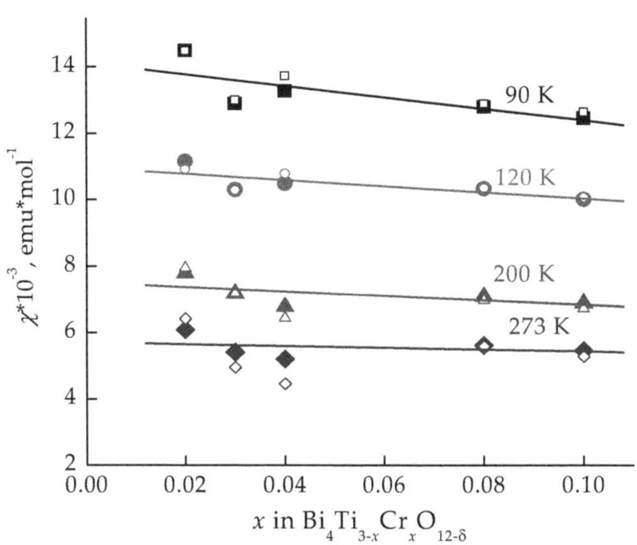

Figure 5.5 Experimental (open symbols) and theoretical (close symbols) paramagnetic components of magnetic susceptibility of $Bi_4Ti_{3-x}Cr_xO_{12-\delta}$.

Acceptable agreement between the experimental magnetic susceptibilities and the calculated values allow us to consider the assumption about the tendency to dimerization of chromium atoms in layered perovskite not contradicting the experiment. The tendencies of aggregation of paramagnetic atoms in perovskite-like mixed oxides were observed previously, in particular, among chromium-containing layered structures. According to the model of exchange channels, in the perovskite layer, superexchange at an angle of 180° is possible via the channels: $d_{xz}|p_x|d_{xz}$ and $d_{yz}|p_y|d_{yz}$ (d_π–p_π interaction).[42,43]

A comparison of magnetic properties of chromium-containing bismuth titanates and studied previously diluted chromium-containing bismuth niobates with layered perovskite structure, $Bi_5Nb_{3-3x}Cr_{3x}O_{15-\delta}$ ($P4/mmm$ for $0 < x \leq 0.04$; $a = 5.47$ Å; $c = 20.97$ Å; $B2/m$ for $x > 0.04$) shows significant differences in their magnetic behavior.[46] In the substituted bismuth niobate the unit cell contains a double layer of perovskite-like blocks separated by bismuth-oxygen layers. Bismuth atoms are located also inside these perovskite blocks. For the solid solutions the exchange parameter was found to be $J = -10$ cm^{-1} and the fraction of dimers a_2 varying from 0.20 ($x = 0.03$) to 0.28 ($x = 0.05$). The effective magnetic moment of Cr^{3+} calculated for the infinite dilution by extrapolating $\chi_{Cr}(x)$ in the solid solutions of bismuth niobate substantially differs from $\mu_{eff}(x \to 0)$ for bismuth titanate both by its value and by its temperature dependence. The effective magnetic moment for $Bi_5Nb_{3-3x}Cr_{3x}O_{15-\delta}$ ($x \to 0$) is 3.40 μB and does not depend on temperature.[46] A certain decrease in μ_{eff} as compared to spin-only value for Cr^{3+} may be associated with the presence of a certain fraction of diamagnetic Cr^{6+}. In the case of $Bi_2BaNb_{2-x}Cr_xO_{9-\delta}$ with layered perovskite-like structure the extrapolation of $\chi_{Cr}(x)$ to the infinite dilution results in $\mu_{eff} = 3.80$ μB, which is close to the spin-only value for single Cr^{3+} atoms.[47] Therefore, antiferromagnetic exchange interactions in chromium-containing solutions of bismuth titanate, $Bi_4Ti_{3-x}Cr_xO_{12-\delta}$ ($J = -25$ cm^{-1}) appear to be stronger than in chromium-containing layered bismuth niobates, $Bi_5Nb_{3-3x}Cr_{3x}O_{15-\delta}$ and $Bi_2BaNb_{2-x}Cr_xO_{9-\delta}$. The exchange-coupled chromium dimers in $Bi_4Ti_{3-x}Cr_xO_{12-\delta}$ remain in noticeable quantities even at the infinite dilution ($a_2(x \to 0) \approx 0.3$). The difference between the exchange parameters may be related to a difference in the angles of exchange. The aggregation of chromium atoms in bismuth titanates may be due to greater covalence of the Cr–O bond in titanate, near a more ionic Ti–O bond, compared to the less ionic Nb–O bond in niobates. An increase in the covalence of the transition element—oxygen bond results in an increase in the aggregation of these atoms. Under heterovalent substitution of chromium atoms for titanium atoms, for maintaining electroneutrality two chromium atoms must be located near a vacancy—$2Cr^{3+}$+[O]. This may be the reason for stable dimerization of chromium atoms and was observed in lanthanum gallate upon heterovalent doping.[48]

5.2.4 Magnetic Properties of Iron-Containing Bismuth Titanates and Niobates

The magnetic behavior of iron-doped layered bismuth titanates has been studied for the $Bi_4Ti_{3-x}Fe_xO_{12-\delta}$ $(0.03 < x \leq 1.33)$ solid solutions, their structure and parameters being given in Table 5.4. The oxidation state of iron atoms being Fe^{3+} in bismuth titanate $Bi_4Ti_3O_{12}$ matrix was previously determined by Mössbauer spectroscopy method.[14]

Table 5.4 Unit cell parameters and space groups for the $Bi_4Ti_{3-x}Fe_xO_{12-\delta}$ (25°C)

Composition	Space group	Unit cell parameters (Å)		
		a	*b*	*c*
$Bi_4Ti_3O_{12}$	*B*2*cb*	5.411	5.448	32.830
$Bi_4Ti_{2.051}Fe_{0.049}O_{12.03}$	*B*2*cb*	5.411	5.447	32.840
$Bi_4Ti_{2.920}Fe_{0.080}O_{12.00}$	*B*2*cb*	5.412	5.448	32.843
$Bi_4Ti_{2.50}Fe_{0.50}O_{11.81}$	*B*2*cb*	5.411	5.451	32.830
$Bi_4Ti_{2.12}Fe_{0.88}O_{11.72}$	*Fmm*2	5.433	5.474	41.194
$Bi_4Ti_{1.67}Fe_{1.33}O_{11.15}$	*Fmm*2	5.427	5.488	41.210

Concentration dependences $\chi(x)$ and temperature dependence of $\mu_{\text{eff}}(T)$ for $Bi_4Ti_{3-x}Fe_xO_{12-\delta}$ and also magnetic moments at infinite dilution $(x \to 0)$ are shown in Figs. 5.6 and 5.7.

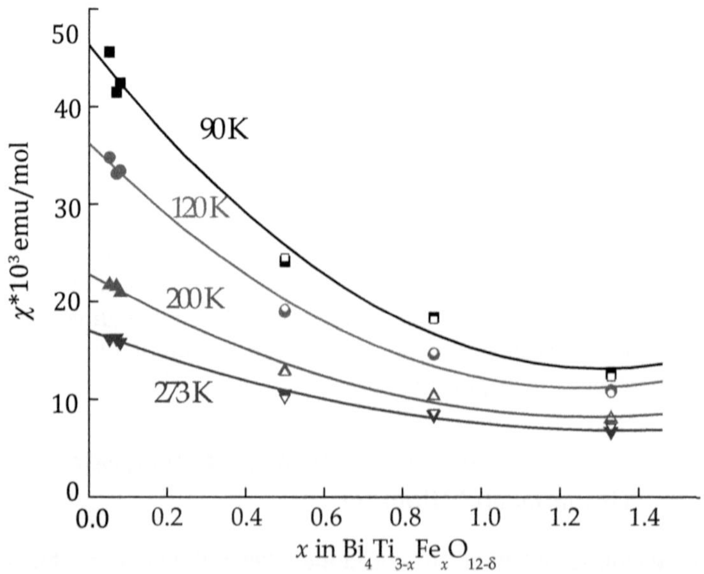

Figure 5.6 Concentration dependences $\chi(x)$ (closed symbols: experimental data, open symbols: calculated values).

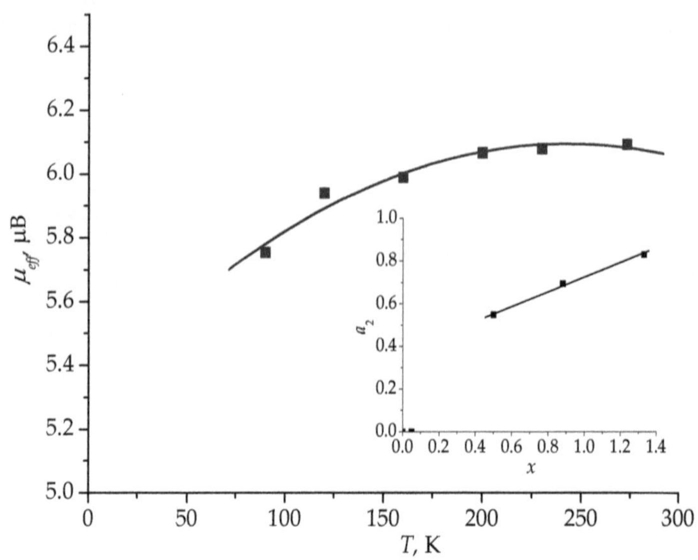

Figure 5.7 Temperature dependence of μ_{eff} at infinite dilution ($x \rightarrow 0$) for $Bi_4Ti_{3-x}Fe_xO_{12-\delta}$. In the insert: fractions of dimers (a_2) in $Bi_4Ti_{3-x}Fe_xO_{12-\delta}$ ($J = -30\ cm^{-1}$).

In the compositions with $x < 0.1$ the values of effective magnetic moment are close to spin-only value (5.92 μB) for Fe^{3+} ($^6A_{1g}$ term) and significantly decrease as the concentration of iron in the solid solutions increases, which is typical for antiferromagnetic character of the exchange interaction.

The inverse paramagnetic component of magnetic susceptibility (χ^{-1}) of $Bi_4Ti_{3-x}Fe_xO_{12-\delta}$ linearly varies with temperature at $T > 100$ K. Magnetically concentrated iron-containing bismuth titanates ($x \geq 0.5$) are characterized by large negative values of Weiss constant. We can expect strong antiferromagnetic interaction and the aggregation of paramagnetic atoms in the solid solutions. These compositions were used to determine the parameter of the exchange interaction and the fraction of Fe^{3+}–O–Fe^{3+} dimers. As the effective magnetic moment of a single iron atom the spin-only value (5.92 μB for d^5) was taken. In the calculation of the magnetic moment of a dimer we used Eq. 1.21 and Table 5.3 (see Chapter 1) for $S_1 = S_2 = 5/2$ and $J = -30$ cm^{-1}. The fraction of dimers, obtained at the best agreement of experimental and calculated magnetic susceptibilities of the solid solution, is shown in Fig. 5.7 (in insert).

It can be seen that the calculation model taking into account only the state of iron as single atoms and dimers results in a good agreement with the experiment, and at $x = 1.3$ (more than 40% substitution) the fraction of dimers is close to 1. Of course, at higher concentrations larger aggregates which formed, for example, from four iron atoms (two ferromagnetic pairs are bonded antiferromagnetically) may already exist. However, to understand the processes of clustering of iron atoms it does not change anything fundamentally.

The isotherms of paramagnetic component of magnetic susceptibility for $Bi_5Nb_{3-3x}Fe_{3x}O_{15-\delta}$ is typical for compounds with antiferromagnetic type of exchange interactions (Fig. 5.8).[49]

The tendencies of iron atoms to aggregate in layered bismuth titanates and niobates can be analyzed by comparing the effective magnetic moments of iron atoms at the infinite dilution ($x \to 0$) obtained by extrapolating the concentration dependences of magnetic susceptibility (Table 5.5) for layered perovskite-like iron-containing bismuth titanate, $Bi_4Ti_{3-x}Fe_xO_{9-\delta}$ and bismuth niobate, $Bi_5Nb_{3-x}Fe_xO_{15-\delta}$. The greater effective magnetic moments compared to the spin-only value in the layered bismuth niobates point to the presence of a certain fraction of ferromagnetically coupled aggregates (dimers with J = 38 cm^{-1}) even at the infinite dilution. When the iron content increases, antiferromagnetic interaction prevails. The aggregation with the formation of dimers is characterized by an exchange parameter J = –55 cm^{-1}.[49]

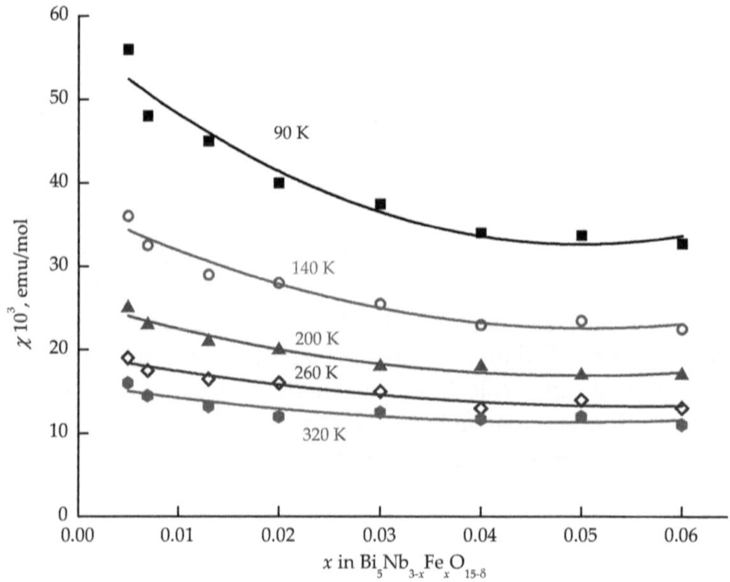

Figure 5.8 Isotherms of paramagnetic component of magnetic susceptibility of iron-containing solid solutions $Bi_5Nb_{3-x}Fe_xO_{15-\delta}$ at 90, 140, 200, 260, and 320 K.[49]

Table 5.5 Effective magnetic moments of iron atom at the infinite dilution

	μ_{eff}(Fe) ($x \to 0$) (μB)			
Composition	**90 K**	**140 K**	**200 K**	**260 K**
$Bi_4Ti_{3-x}Fe_xO_{9-\delta}$[45]	5.88	5.88	5.95	6.00
$Bi_5Fe_xNb_{3-x}O_{15-\delta}$[49]	6.87	6.88	6.90	6.91

Thus, the dilution of layered iron-containing bismuth titanates results in a complete disaggregation of iron atoms. In layered bismuth niobates, even at the infinite dilution, ferromagnetic dimers along with single iron atoms seem to be preserved. With the increase of iron content, a sufficiently strong antiferromagnetic exchange manifests itself both in layered perovskite bismuth titanates (J = −30 cm^{-1}) and in bismuth niobates (J = −55 cm^{-1}). Clustering of iron atoms in the form of antiferromagnetically coupled dimers or more complex aggregates in the solid solutions based on bismuth niobates is more pronounced.

5.2.5 Magnetic Properties of Manganese Containing Bismuth Titanates and Niobates

The structural characteristics of the synthesized solid solutions $Bi_4Mn_xTi_{3-x}O_{12-\delta}$ with layered perovskite structure are given in Table 5.6.

Table 5.6 Space groups and unit cell parameters of $Bi_4Ti_{3-x}Mn_xO_{12-\delta}$ (25°C)

		Unit cell parameters (Å)		
Composition	**Space group**	***a***	***b***	***c***
$Bi_4Ti_{2.90}Mn_{0.10}O_{11.7}$	*B2cb*	5.406	5.418	32.791
$Bi_4Ti_{2.70}Mn_{0.30}O_{11.7}$	*B2cb*	5.400	5.414	32.773
$Bi_4Ti_{2.60}Mn_{0.40}O_{11.6}$	*B2cb*	5.394	5.414	32.740
$Bi_4Ti_{2.40}Mn_{0.60}O_{11.4}$	*B2cb*	5.402	5.416	32.733

According to NEXAFS spectra of $Bi_4Mn_{0.5}Ti_{2.5}O_{12-\delta}$ (Fig. 5.9), manganese in bismuth titanates solid solutions with layered perovskite structure is mainly in the form of Mn^{3+} and Mn^{4+}, the spin-only values of the effective magnetic moment for these states being 4.90 and 3.87 μB, respectively.

The concentration dependences of magnetic susceptibility of manganese-containing bismuth titanate $Bi_4Mn_xTi_{3-x}O_{12-\delta}$ with layered perovskite structure is given in Fig. 5.10. The temperature dependences of magnetic moment μ_{eff}(Mn) of $Bi_4Mn_xTi_{3-x}O_{12-\delta}$ on concentration and for the infinite dilution ($x \to 0$) are given in Fig. 5.11.

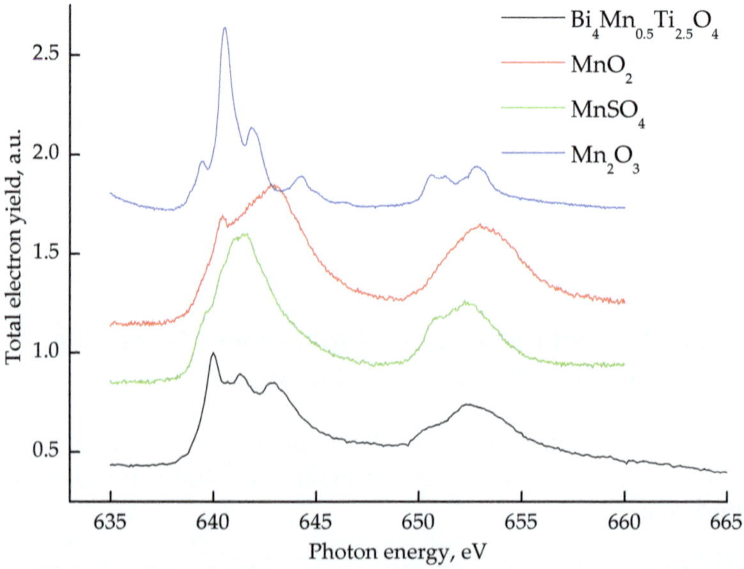

Figure 5.9 NEXAFS Mn2*p*-spectra of $Bi_4Mn_{0.5}Ti_{2.5}O_{12-\delta}$, and of MnO_2, Mn_2O_3, $MnSO_4$ for comparison.

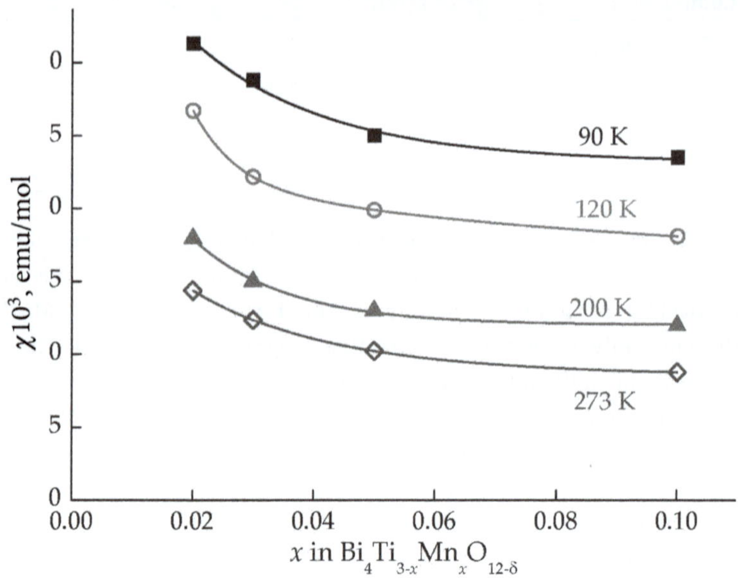

Figure 5.10 Concentration dependences of magnetic susceptibility $\chi(x)$ for $Bi_4Mn_xTi_{3-x}O_{12-\delta}$.

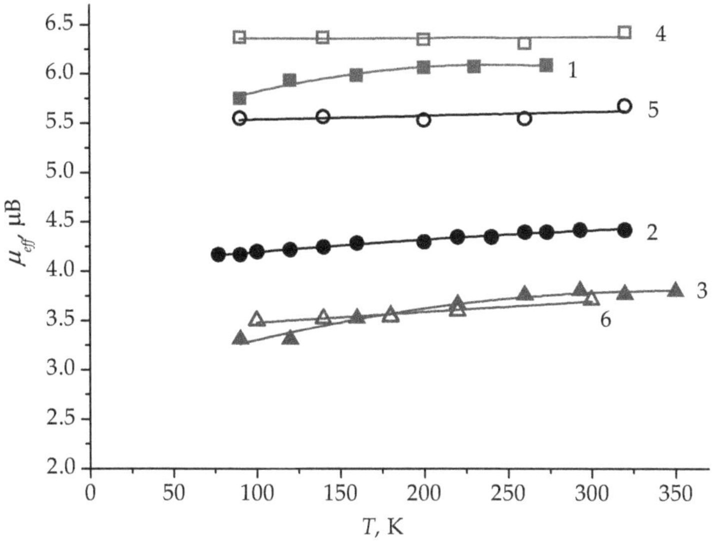

Figure 5.11 Temperature dependence of the effective magnetic moment of paramagnetic atoms in $Bi_4Ti_{3-x}M_xO_{12-\delta}$ (*M*—Fe (1), Mn (2), Cr (3)) and in $Bi_5Nb_{3-3x}M_{3x}O_{15-\delta}$ (*M*—Fe (4), Mn (5), Cr (6)).

The effective magnetic moment of manganese obtained by extrapolating $\chi_{Mn}(x)$ to the infinite dilution as well as experimentally determined values for synthesized solid solutions noticeably increase as the temperature increases. In addition, there is a decrease in the magnetic moment in solid solutions from 5.72 μB ($x = 0.02$) to 4.2 ($x = 0.1$) at 298 K and 4.8 to 4.2 μB at 90 K, when the content of manganese increases.

As for the magnetic behavior of $Bi_4Mn_xTi_{3-x}O_{12-\delta}$ solid solutions, the shapes of the concentration dependences of the magnetic susceptibility point to antiferromagnetic exchange interactions between paramagnetic atoms, as it could be expected for the structures including perovskite-like layers (Fig. 5.10).

However, the magnetic moments for the most diluted solutions (lowest concentrations of manganese) and for the infinitely diluted solution distinctly exceed the magnetic moments for Mn^{4+}, Mn^{3+} and their combination. There are two ways to account for this behavior. First, we could assume the presence of Mn^{2+} with μ_{eff} = 5.92 μB in significant quantities (>40%), which is unlikely and is not confirmed by NEXAFS spectroscopy.

The more probable cause of the increased value of the magnetic moment may be the formation of ferromagnetically coupled clusters of neighboring manganese atoms. Then, in manganese-containing bismuth titanates with layered perovskite structure two types of exchange interactions between manganese atoms are realized: ferromagnetic interaction, which is due to the presence of manganese in different oxidation states—Mn^{4+} and Mn^{3+}—and antiferromagnetic 180° exchange.

The estimation of the state of manganese atoms in the solid solution at the infinite dilution with the assumption of the presence of single Mn^{3+} and dimers Mn^{3+}–O–Mn^{3+} or Mn^{2+}–O–Mn^{4+} which result from disproportionation of Mn^{3+} gives ferromagnetic exchange parameter J = 40 cm^{-1} (fraction a_2 = 0.5). The nonlinear dependence of magnetization of the sample on the field at 4 K (Fig. 5.12) points to a ferromagnetic contribution in magnetic moment.

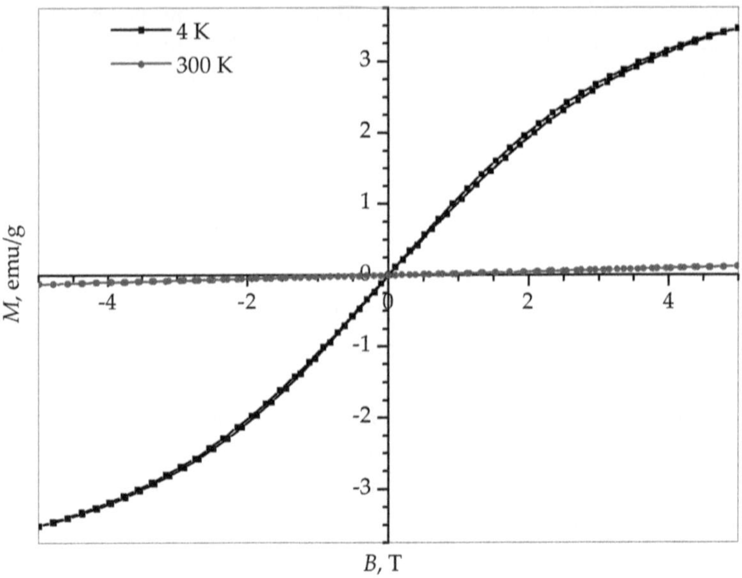

Figure 5.12 The dependence of specific magnetization of $Bi_4Ti_{2.95}Mn_{0.05}O_{12-\delta}$ on the field at 4 and 300 K.

The isotherms of paramagnetic component of magnetic susceptibility for $Bi_5Nb_{3-3x}Mn_{3x}O_{15-\delta}$ is typical for compounds with antiferromagnetic type of exchange interactions (Fig. 5.13).[50]

The effective magnetic moments of manganese atoms, calculated from extrapolation of the concentration dependences of $\chi(Mn)$ to the infinite dilution of solid solutions increase as the temperature increases from $\mu_{eff}(Mn)$ = 5.72 μB (T = 90 K) to 5.86 μB (T = 320 K).

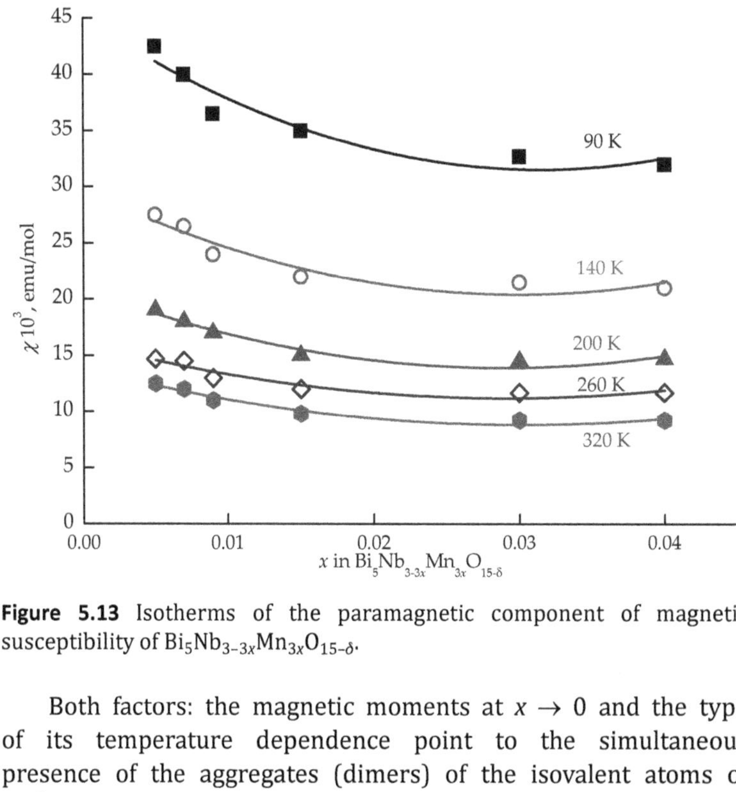

Figure 5.13 Isotherms of the paramagnetic component of magnetic susceptibility of $Bi_5Nb_{3-3x}Mn_{3x}O_{15-\delta}$.

Both factors: the magnetic moments at $x \rightarrow 0$ and the type of its temperature dependence point to the simultaneous presence of the aggregates (dimers) of the isovalent atoms of Mn^{3+} with antiferromagnetic exchange parameter J = –45 cm^{-1}, and a certain quantity of ferromagnetically coupled dimers of Mn^{2+} and Mn^{4+} atoms due to disproportionation of Mn^{3+} with J = 38 cm^{-1}, along with single Mn^{3+} atoms.[50]

Thus, in doped bismuth titanates and niobates with layered perovskite-type structure, as the concentration of the paramagnetic atom increases, an antiferromagnetic exchange is manifested due to the overlapping of orbitals of metal and oxygen atoms at an angle of 180° as the main tendency. When the dopant concentration decreases up to infinite dilution (Fig. 5.13), individual features of paramagnetic atoms begin to show themselves:

ferromagnetic and antiferromagnetic superexchange, clustering (the formation of dimers).

5.3 Magnetic Behavior of Doped Bismuth Titanates and Niobates with Pyrochlore-Type Structure

5.3.1 Pyrochlore Structure Type and Special Features of Bismuth-Containing Pyrochlores

The pyrochlore structure is realized for a large number of binary oxides $A_2B_2X_6Y$ (X, Y = O^{2-}, OH^-, Hal^-). In the space group of cubic pyrochlore $Fd\bar{3}m$, Z = 8 (№227, origin at 16c, 3m) all the atoms in the unit cell occupy specific positions: A (16c), B (16d), X (48f), Y (8a). The atomic arrangement is completely specified, except for the x coordinate of the 48f positions. The principal interatomic distances can be expressed as functions of the atom coordinate x and the cell edge a.[51-54] Geometric analysis for the pyrochlore structure type, relationships between interatomic distances, the values of the anion positional parameter x, the cell edge a and the ratio of ionic radii of A and B cations were considered by Chakoumakos.[54] Varying the x coordinate changes the shapes of A- and B-site polyhedra. For x = 0.375, the anions X are arranged as in anion-deficient fluorite. At this value of x, the oxygen polyhedra surrounding A and B cations (A- and B-site polyhedra) are regular cubes and trigonally flattened octahedra, respectively. When the x-coordinate increases to 0.4375 the B-site becomes a regular octahedron and the cubic A-site is distorted into a trigonal scalenohedron. With further increase in x, the B-site octahedra elongate along a threefold axis and the A-site becomes a hexagonal bipyramid. In order to determine the degree of individual influence of A and B cations on the cell parameter, a regression analysis of the dependence of $x(r_A)$ using ionic radii by Shannon was performed for a large number of pyrochlores.[51] The influence of the ionic radii of A and B cations on the cell parameter a was shown to be almost equal. This result supports the description of the pyrochlore structure as an interpenetration of two sublattices (B_2X_6 and YA_2 or B_2O_6

and A_2O' in the case of oxide pyrochlores) rather than as relatively stable B_2X_6 framework with the A and Y ions filling interstices (Fig. 5.14).[57]

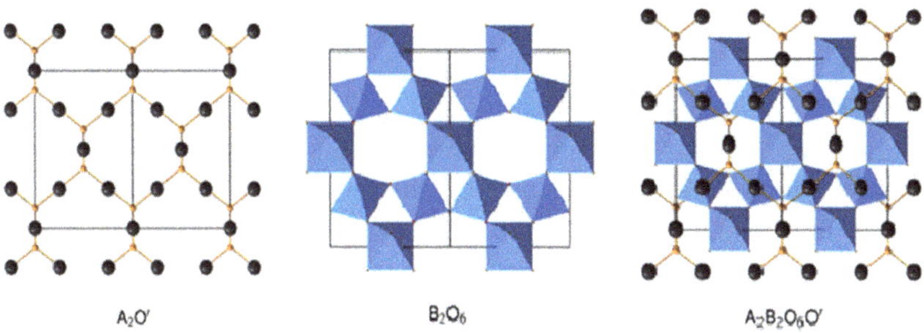

Figure 5.14 The pyrochlore structure in the form of two sublattices A_2O' and B_2O_6.

A disordering in the anionic sublattice (X, Y) also does not affect the parameters a and x. The inconsistencies between the calculated and experimental a values for pyrochlores containing Na^+, Pb^{2+}, and Bi^{3+} ions among A cations are observed most often. This is due to differences in the bond lengths predicted on the basis of ionic radii and experimentally determined, (owing to the distortion of polyhedra, a partial occupancy of the sites, a covalent or metallic bond type. The dependences of the unit cell parameter a for a number of rare-earth metals pyrochlores on ionic radii of A and B cations are shown in Fig. 5.15.

There is a steady tendency to an increase in the unit cell parameter (a) with an increase in the size of the cation. To estimate the possibility of pyrochlore structure formation in binary oxides a pyrochlore structure field map has been contoured for reasonable values of cell edge a, the anion position parameter x and octahedral cation radii r_A/r_B, and the concept of "pyrochlore stability field" was suggested.

The range of the r_A/r_B ratio within 1.46–1.78 limits the formation of pyrochlores at atmospheric pressure. The mentioned range of r_A/r_B determines the formation of the largest number of different mixed oxides with the pyrochlore structure. Large r_A/r_B characterizes pyrochlores, which were obtained at high pressure.[53] For rare earth elements, r_A/r_B varies between 1.61 ($Lu_2Ti_2O_7$) and 1.78 ($Sm_2Ti_2O_7$). The larger atoms of rare earth elements (Nd–La) do not form titanates with the pyrochlore structure. For pyrochlores containing bismuth in *A*-sites and niobium or titanium in *B*-sites, the evaluation of the structural parameters results in the following unit cell parameters and the position parameters *x*: for Nb a = 10.70 Å, x = 0.427 Å; for Ti a = 10.55 Å, x = 0.430. The ionic radii $r(Nb^{5+})_{VI}$ = 0.64 Å; $r(Ti^{4+})_{VI}$ = 0.61 Å; $r(Bi^{+3})_{VIII}$ = 1.17 Å are used.[55] The r_A/r_B is 1.83 and 1.93 for niobium and titanium, respectively. The $Bi_2Ti_2O_7$ with r_A/r_B = 1.93 falls out of "pyrochlore stability field", although the $Bi_{1.85}Rh_2O_{6.83}$ (r_A/r_B = 1.95) is formed as pyrochlore.[74] The ratio r_{Bi}/r_{Nb} = 1.83 is a borderline and bismuth niobate crystallizes in two orthorhombic (low- and high-temperature) modifications.

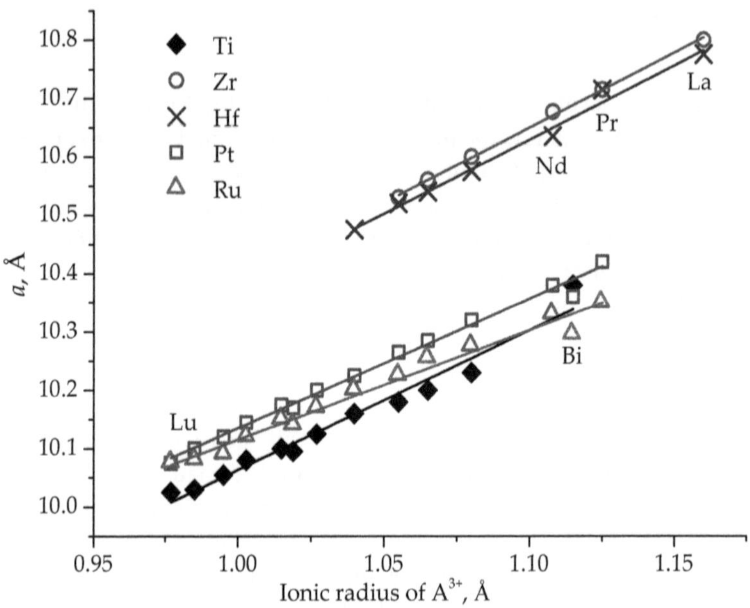

Figure 5.15 The dependences of the unit cell parameter a for $A_2B_2O_6O'$ pyrochlores (*A*—rare-earth metals and bismuth) on ionic radii r_A (298 K).[55,71]

The formation of mixed bismuth niobate with the pyrochlore structure in triple Bi_2O_3–Nb_2O_5–$MO(M_2O_3)$ systems, where *M* is the element having a lower ionic radius than bismuth was discussed in.[52,69] Recently the chromium-, zinc-, iron-, manganese-, copper-containing complex bismuth niobates pyrochlores were obtained[56,58–62] and actively studied (Fig. 5.16). The main efforts of researchers were aimed at the studies of the phase diagrams of triple systems and determining the field of the pyrochlore phase existence, at the neutron diffraction studies of the structure, the investigation of dielectric properties of the pyrochlores.[58–69]

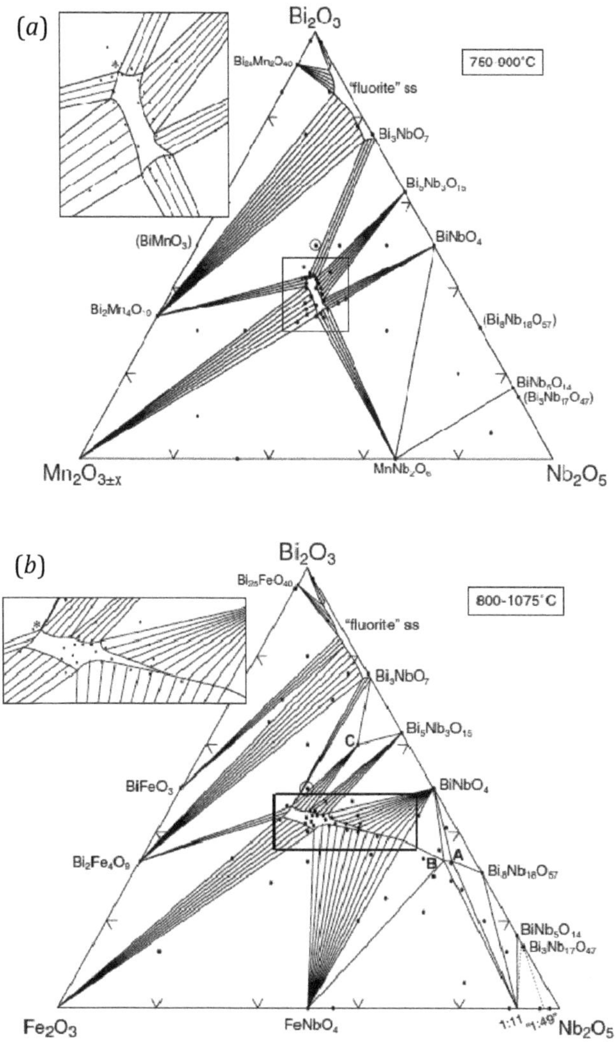

Figure 5.16 Subsolidus phase equilibrium diagrams in air for the systems: Bi_2O_3–$Mn_2O_{3\pm x}$–Nb_2O_5 (*a*),[65] Bi_2O_3–Fe_2O_3–Nb_2O_5 (*b*).[66] Large pyrochlore phase fields are formed near the center of the diagrams.

Examination of phase diagrams results in the following conclusions. A large pyrochlore phase field is formed near the center of the diagram. The "pyrochlore phase fields" in size and location in the phase diagram vary significantly for systems with different substituent atoms (Fig. 5.16). The chromium and copper-containing bismuth niobates pyrochlores are represented by one composition. The formation of stable nickel- and zinc-containing bismuth niobates pyrochlores is observed over a small range of compositions, whereas the largest "fields of pyrochlore" are recorded in manganese and iron-containing compounds. The possibility is found of formation of a large number of solid solutions which retain the pyrochlore structure with bismuth and niobium oxides.

The formation of triple bismuth niobates with pyrochlore structure in a fairly wide range of metals molar ratios was responsible for a series of structural studies with the aim of revealing the cation distribution over the obtained pyrochlores on the basis of neutron diffraction data.[65–69] As a result of these studies, two types of disorder in the substituted bismuth niobates with pyrochlore structure were revealed: structural disordering, when *B*-type cations occupy *A*-sites and displacement of *A* and O′ atoms from the center of the sites of the ideal pyrochlore structure (Table 5.7).

A large group of mixed oxides with pyrochlore structure is known, where bismuth is in the *A*-sites. Kennedy considered in detail the changes occurring in the effective ionic radius of Bi^{3+} in the pyrochlore-insulators and the pyrochlores exhibiting metallic properties.[71] In metallic pyrochlores (*B* = Rh, Ru, Ir, Pt) the cell parameter *a* is systematically smaller than the value predicted from the ionic radii of Pt and Ru. On the contrary, in the $Bi_2Ti_2O_7$ and α-$Bi_2Sn_2O_7$ insulators the cell parameter appears to be much larger than expected (α-$Bi_2Sn_2O_7$ is a tetragonal phase at room temperature and the effective cubic parameter was calculated). For $Bi_2Hf_2O_7$ the cell parameter (a = 10.86 Å) is much higher than expected and is not shown in Fig. 5.15. Also, the displacements of bismuth cations from the ideal pyrochlore positions (16*d*) were found. The distortion of pyrochlore structure of α,β-$Bi_2Sn_2O_7$ and $Bi_2Hf_2O_7$ may be the result of ordering due to a large displacement of bismuth cations (a large displacement of bismuth cations in α,β-$Bi_2Sn_2O_7$ and $Bi_2Hf_2O_7$ may result in the distortion of pyrochlore structure).

Table 5.7 Structural parameters of the substituted bismuth niobates pyrochlores

Composition	a(Å)	x(48f)	δBi(x,y) 96g/96h	δO′ (x,y,z) (32e)	Occ. (A); (B) Bi/M; Nb/M	Occ. O/O′
$Bi_{1.6}Mn_{1.2}Nb_{1.2}O_7$[65]	10.47616	0.3206	0.361 (96g)	0.33	0.80/0.20(A) 0.60/0.40(B)	1/0.08
$Bi_{1.72}Fe_{1.07}Nb_{1.13}O_7$ $Bi_{1.721}Fe_{0.190}$ $(Fe_{0.866}Nb_{1.134})O_7$[66]	10.5036 (10 K) 10.5103 (298 K)	0.3197	0.430 (96g)	0.29	0.860/0.016 (A) 0.433/0.567 (B)	1/0.23
$Bi_{1.7}CoNb_2O_{7-\delta}$[67]			0.400 (96g)		0.80/0.18(A) 0.60/0.40(B)	
$Bi_{1.67}Ni_{0.75}Nb_{1.50}$ O_7[68]	10.5354	0.3199	0.428 (96h)	0.393	0.83/0.12(A) 0.75/0.25(B)	1/0.99
$Bi_{1.67}Mg_{0.64}Nb_{1.53}$ O_7[68]	10.5607	0.3198	0.415 (96h)	0.370	0.84/0.06(A) 0.75/0.25(B)	1/0.97
$Bi_{1.51}Zn_{0.96}Nb_{1.49}$ $O_{6.96}$ $Bi_{1.54}Zn_{0.83}Nb_{1.54}$ O_7[63,60]	10.5465 10.56339	0.3196	0.400 (96g)	0.39	0.75/0.16(A) 0.75/0.24(B)	1/0.8 0.97/0.97
$Bi_2Ti_2O_7$[70]	10.379 10.335	0.4315	0.390 (96g)	0.46	1	1

The changes in the cell parameter may be associated with ionic radius of Bi^{3+}. The decrease in the radius of Bi^{3+} in metallic pyrochlores is explained by the decrease in the influence of the lone electron 6s pair due to the overlapping of 6s(Bi)–d(B) orbitals.

A detailed neutron diffraction study of $Bi_2Ti_2O_7$ pyrochlore structure using various simulation methods is presented by A. Hector and his colleagues.[70] It is noted that the lattice parameter of $Bi_2Ti_2O_7$ was significantly larger than previously reported.[72] The results of the study show a high mobility of bismuth and oxygen atoms in the Bi_4O' sublattice and their displacements from the ideal pyrochlore sites due to the effect of the lone s-electron pair (Fig. 5.17b).

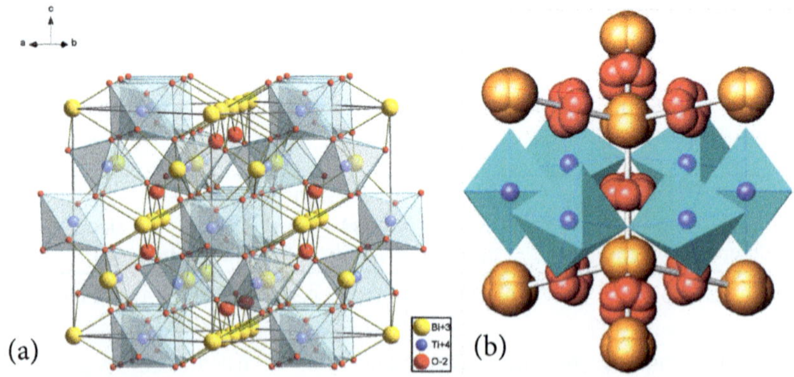

Figure 5.17 The ideal pyrochlore structure of $Bi_2Ti_2O_6O'$ (a) and the fragment of bismuth surrounding taking into account the displacement of bismuth atoms from the 16*d* ideal pyrochlore sites to 96*g* (the sixfold static disorder, red spheres) and the displacement of O′ atoms from the 8*a* sites to the 32*e* sites (the fourfold disorder, light brown spheres) (b). In the center of oxygen octahedra (48*f*) are titanium atoms.[70]

The structure of stoichiometric $Bi_2Ti_2O_7$ and the polyhedra of cations suggested by the authors are given in Fig. 5.17.

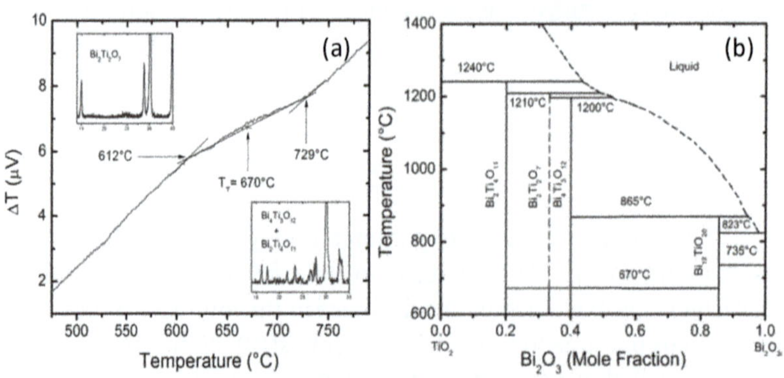

Figure 5.18 DTA curves of the $Bi_2Ti_2O_7$ powder. In the inserts: the X-ray patterns of $Bi_2Ti_2O_7$ at temperatures below 612°C (left) and above 729°C (a).[73] Phase diagram of the Bi_2O_3-TiO_2 system. The vertical dotted line for $Bi_2Ti_2O_7$ shows an unstable state (b).[73]

Structural disordering of $Bi_2Ti_2O_7$, along with dimensional (geometric) factor (r_A/r_B = 1.93—falls out of the structure stability region) results in the thermal instability of the bismuth titanate. The unstable phase of the $Bi_2Ti_2O_7$ pyrochlore, which is decomposed at temperatures above 670°C to give $Bi_2Ti_4O_{11}$ and $Bi_4Ti_3O_{12}$ is shown in the phase diagram of the Bi_2O_3-TiO_2 system (Fig. 5.18).[73] Doping bismuth titanate with 3*d*- and rare earth metals increases their thermal (thermodynamic) stability.

5.3.2 Magnetic Properties of Chromium, Iron, Manganese Containing Bismuth Niobates and Titanates with Pyrochlore-Type Structure

For bismuth niobates containing iron or manganese in the molar ratios comparable with niobium and bismuth and characterized by pyrochlore-type structure the study of magnetic behavior was carried out earlier and antiferromagnetic exchange interactions between paramagnetic atoms were found.[65,66] There is no information concerning the magnetic behavior of multicomponent (doped with 3*d*-elements) bismuth titanates with pyrochlore structure. It is obvious that the magnetic behavior of multi-component niobates and bismuth titanate (pyrochlore) is determined by their composition, the local position and electronic structure of paramagnetic atoms. Therefore, a detailed study of the structure and the state of atoms of the doping element is required, including the determination of cations distribution and the oxidation states of the paramagnetic atoms. Below we report the results of the study of magnetic behavior of chromium-, manganese-, and iron-doped titanates and niobates of bismuth with pyrochlore structure. The single-phase compounds and solid solutions were obtained by solid-phase method, certified by the method of DTA, X-ray diffraction analysis, and energy dispersive electron-microscopic analysis. The oxidation state of paramagnetic atoms was determined by EPR, Mössbauer, and EXAFS spectroscopy.

The study of cation distribution in doped bismuth titanates was carried out by X-ray powder diffraction study of synthesized solid solutions. The pycnometric densities and the densities calculated for various models of the doping element distribution were compared. The Rietveld refinement of the structure of $Bi_{1.6}M_xTi_2O_{7-\delta}$ (M = Cr, Mn, Fe) using X-ray data was carried out using FullProf software.[74] The occupancies of the atoms sites were fixed according to the determined quantitative content of the elements. We considered several models of the distribution of the dopant over *A*- and *B*-sites in the pyrochlore structure: the ideal pyrochlore structure and (1) all the doping atoms are distributed over *A*-sites; (2) a quarter of the doping atoms are distributed over the *B*-sites, and three quarters over *A*-sites; (3) the doping atoms are distributed over *A*- and *B*-sites in the same molar fraction (0.5).

5.3.2.1 Magnetic properties of $Bi_{1.6}Cr_xTi_2O_{7-\delta}$ and $Bi_{1.6}Mg_{0.8-x}Cr_xNb_{1.6}O_{7-\delta}$

The magnetic behavior of the $Bi_{1.6}Cr_xTi_2O_{7-\delta}$ (x = 0.016–0.091) and $Bi_{1.6}Mg_{0.8-x}Cr_xNb_{1.6}O_{7-\delta}$ with pyrochlore structure was studied.[75,76] Synthesized solid solutions were investigated by DSC, and radiographic methods previously. The observed, calculated for the optimal model of the distribution of chromium atoms over cationic sites and difference X-ray powder diffraction pattern for $Bi_{1.6}Cr_{0.16}Ti_2O_{7-\delta}$ are given in Fig. 5.19 and Table 5.8.

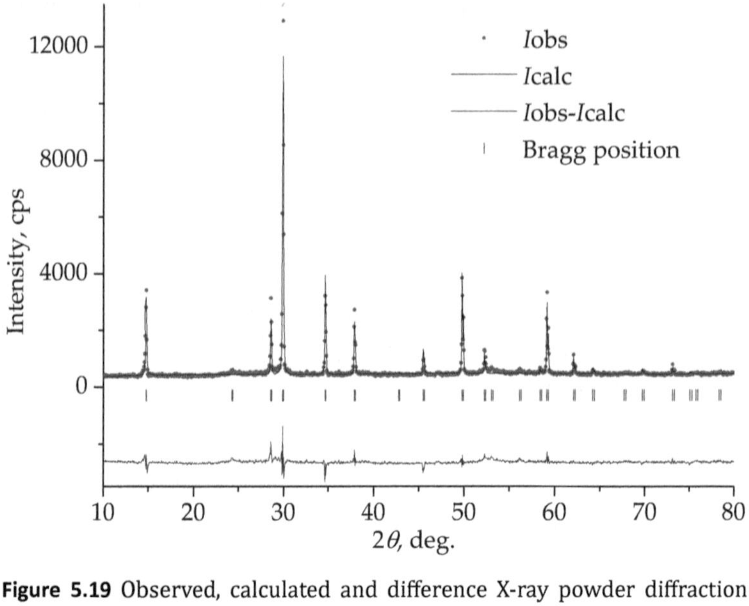

Figure 5.19 Observed, calculated and difference X-ray powder diffraction patterns of $Bi_{1.6}Cr_{0.16}Ti_2O_{7-\delta}$.

Table 5.8 Atomic coordinates (*x*, *y*, *z*), anisotropic thermal parameters (U_{11}, U_{22}, U_{33}), and occupancy in $Bi_{1.6}Cr_{0.16}Ti_2O_{6.64}$ by Rietveld refinement (R_p = 7.20%; R_{wp} = 9.37%; χ^2 = 4.76)

Atom	Site	*x*	*y*	*z*	U_{11}, U_{22}	U_{33}	Occupancy
Bi/Cr	16*c*	0	0	0	0.00052(4)	0.00052(4)	0.8/0.08
Ti/Cr	16*d*	1/2	1/2	1/2	0.0045(3)	0.0045(3)	1/0
O	48*f*	1/8	1/8	0.424(6)	0.0095(3)	0.0182(3)	1
O′	8*a*	1/8	1/8	1/8	0.1	0.1	0.33

Note: Anisotropic thermal parameters units Å × 100, *a* = 10.3486(7) Å.

NEXAFS Cr2*p* spectra of the $Bi_{1.6}Cr_{0.16}Ti_2O_{6.595}$ and several chromium compounds in stable oxidation states (for comparison) are given in Fig. 5.20.

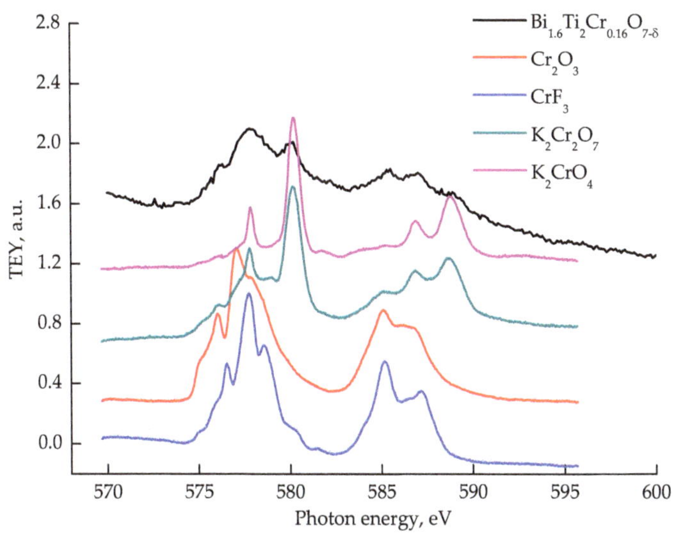

Figure 5.20 NEXAFS Cr2p spectra of the $Bi_{1.6}Cr_{0.16}Ti_2O_{6.59}$, chromium(III) oxide, chromium(III) fluoride, potassium chromate and dichromate.

A comparison of the shape and the energies of the lines of the fine structure of the absorption band of chromium-containing bismuth titanate with the pyrochlore structure and of the chromium(III) oxide and fluoride, and potassium chromates allows us to conclude that the major part of chromium atoms is in the form of Cr^{3+}.

Concentration dependence of magnetic susceptibility and the temperature dependence of the magnetic moment of chromium in $Bi_{1.6}Cr_xTi_2O_{7-\delta}$ at infinite dilution ($x \rightarrow 0$) are given in Figs. 5.21 and 5.22.

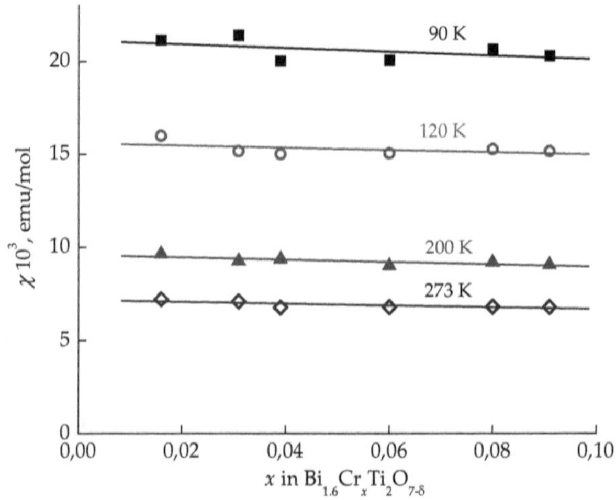

Figure 5.21 Dependence of magnetic susceptibility on the concentration of chromium in $Bi_{1.6}Cr_xTi_2O_{7-\delta}$.

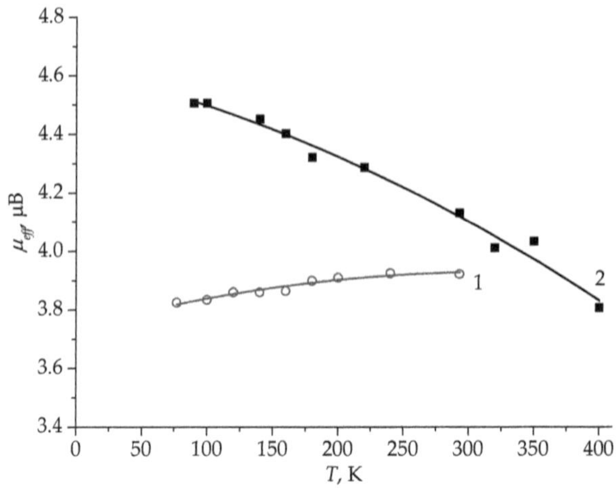

Figure 5.22 Temperature dependence of the effective magnetic moment at the infinite dilution in $Bi_{1.6}Cr_xTi_2O_{7-\delta}$ (1) and $Bi_{1.6}Mg_{0.8(1-x)}Cr_{0.8x}$ $Nb_2O_{7-\delta}$ (2).

The dependence of $\chi^{-1}(T)$ in $Bi_{1.6}Cr_xTi_2O_{7-\delta}$ obeys Curie's law. The intersection points (x) of the $\chi^{-1}(T)$ plots with the temperature axis slightly deviate from 0 in both directions (-7 K $\leq x \leq$ 3.4 K). The effective magnetic moments slightly increase (3.8 μB $\leq \mu_{eff} \leq$ 3.9 μB) and are close to spin-only effective magnetic moment of Cr^{3+} (3.87 μB) in the studied concentration and temperature range, which points to the absence of exchange interactions in chromium-containing bismuth titanate (pyrochlores).[75] Obviously, the distribution of chromium in the obtained solid solutions over the bismuth sites does not contribute to exchange interaction due to the absence of channels for overlapping Cr^{3+}–O–Cr^{3+} orbitals.

Chromium-containing bismuth niobates with the pyrochlore-type structure are formed only at high chromium content, for example, in $Bi_{1.34}CrNbO_{7-\delta}$.[57] We synthesized single-phase $Bi_{1.6}Cr_{0.8}Nb_{1.6}O_{7-\delta}$ with the pyrochlore-type structure (a = 10.451(3) Å).

The dependences of magnetization of the sample on the field at 2, 30 and 300 K are given in the inserts in Fig. 5.23. The data obtained in ZFC and FC (0.05 T) differ insignificantly. The ferromagnetic contribution at low temperatures is confirmed by the nonlinear dependence of magnetization of the sample on the field at 2 K. When the sample is heated to 30 K and more, the magnetization decreases and the changes with the field increase linearly. The dependences of magnetic susceptibility and their inverse values on the temperature are given Fig. 5.23. The temperature dependence of the inverse paramagnetic component of magnetic susceptibility is described by two linear intervals, which obey the Curie–Weiss law with Weiss constants –3.6 K at $T \leq$ 150 K and –153.6 K at higher temperatures. The effective magnetic moment increases from 2.5 to 3.5 μB in the temperature range from 2 K to 20 K, remains constant (3.5) μB in the range 20 K $\leq T \leq$ 100 K, and increases to 4.1 μB at 300 K. Antiferromagnetic interaction exists all over the lattice and becomes cooperative in the samples with high chromium content and the distribution of chromium over B-sites. The ferromagnetic exchange can exist in the aggregates containing two neighboring atoms located in different sublattices—A (bismuth) and B (niobium). Such aggregates may determine an increase in the effective magnetic moment in comparison with the spin-only value after heating the sample at $T >$ 100 K.

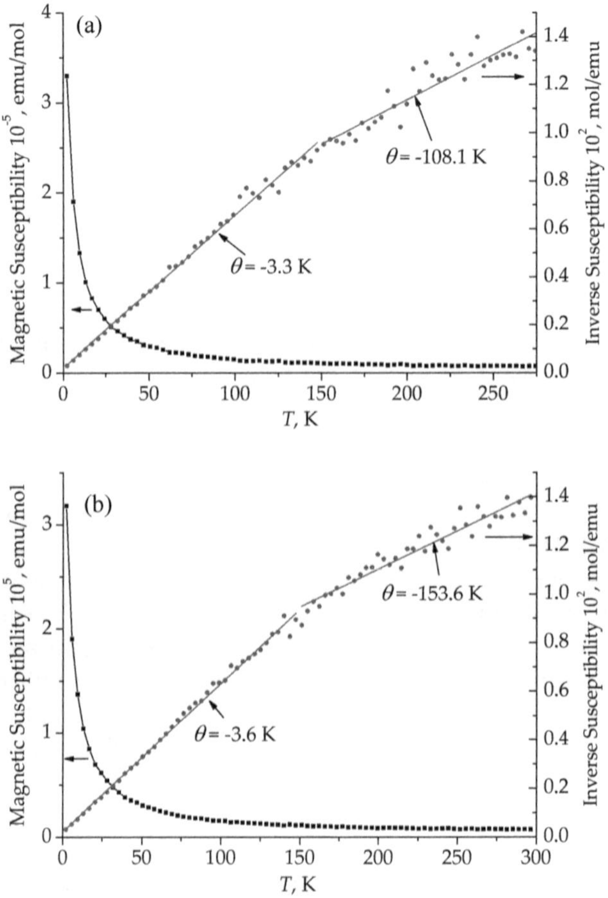

Figure 5.23 Temperature dependencies of magnetic susceptibility and their inverse values of $Bi_{1.6}Cr_{0.8}Nb_{1.6}O_{7-\delta}$ at FC = 0.05 T (a), ZFC (b).

Diluted solid solutions, in which it is possible to study the state of paramagnetic atoms, not complicated by cooperative interactions, can be obtained only by stabilizing the pyrochlore structure by introduction of other diamagnetic atoms. So, both magnesium and chromium-containing bismuth niobates with varying content of these elements were synthesized and their magnetic properties were investigated.[75] The concentration dependence of magnetic susceptibility and the temperature dependence of the magnetic moment of chromium in $Bi_2Mg_{1-x}Cr_xNb_2O_{7+\delta}$ ($Bi_{1.6}Mg_{0.8(1-x)}Cr_{0.8x}Nb_2O_{7-\delta}$ as a formula unit (FU) for pyrochlore structure) at the infinite dilution ($x \rightarrow 0$) are given in Fig. 5.25.

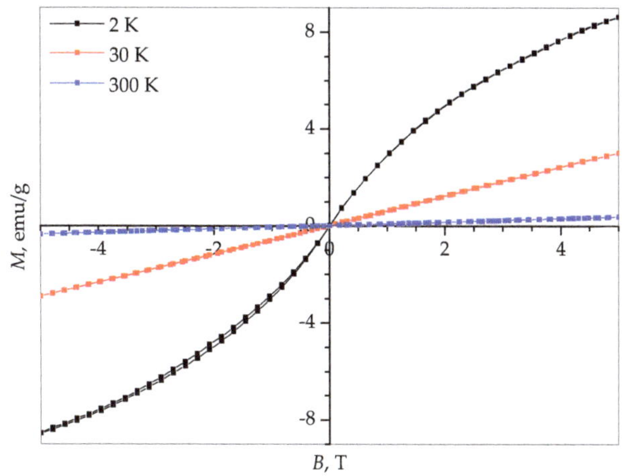

Figure 5.24 The field dependences of specific magnetization of $Bi_{1.6}Cr_{0.8}Nb_{1.6}O_{7-\delta}$ at 2, 30, and 300 K.

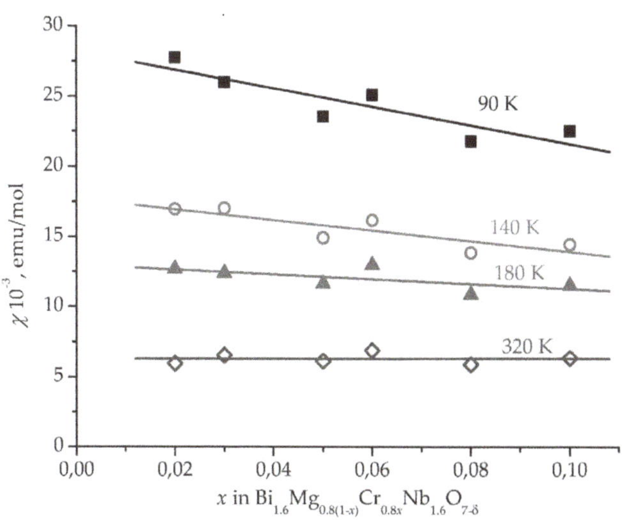

Figure 5.25 Dependence of magnetic susceptibility on the concentration of chromium in $Bi_{1.6}Mg_{0.8(1-x)}Cr_{0.8x}Nb_2O_{7-\delta}$.

Magnetic susceptibility and the effective magnetic moment decreases with increasing chromium content in the solid solutions. Extrapolation of magnetic susceptibility to the infinite dilution results in rather large μ_{eff} (up to 4.7 μB at $T \leq 100$ K), which can be explained only from the standpoint of the formation of exchange-coupled dimers with an exchange parameter $J = 18\ \text{cm}^{-1}$ and the fraction of $a_2 = 0.48$ calculated by the above-mentioned method (see Chapter 1).

The presence of clusters with ferromagnetic interaction at infinite dilution of the solid solution can be associated with the emergence of ferromagnetic exchange correlations between neighboring chromium atoms located in different equivalent positions in the sublattices of bismuth and niobium, which are leveled by antiferromagnetic interaction as the chromium content increases and the chromium atoms become distributed over niobium sites. In mixed bismuth niobate containing both chromium and magnesium, the atoms of these elements can be distributed over the sites of bismuth and niobium because both boundary compositions—$Bi_{1.34}CrNbO_{7-\delta}$[57] and $Bi_2MgNb_2O_9$ (FU—$Bi_{1.6}Mg_{0.8}Nb_{1.6}O_{7-\delta}$)—are stable.[68] In the vicinity of magnesium atoms additional vacancies are formed, which can contribute to the clustering of chromium atoms, since magnesium gives more ionic bonds with oxygen, which results in the increase in the covalence of the bond of the nearest chromium atom with oxygen and also in the increase in the clustering of chromium atoms in bismuth niobates.

Thus, in chromium-containing bismuth titanates $Bi_{1.6}Cr_xTi_2O_{7-\delta}$ chromium atoms are distributed only over *A*-sites of the pyrochlore structure and do not participate in exchange interactions. In diluted chromium-containing bismuth niobate $Bi_{1.6}Mg_{0.8(1-x)}Cr_{0.8x}Nb_{1.6}O_{7-\delta}$ chromium atoms, which can occupy both *A*- and *B*-sites show the ferromagnetic interaction with an exchange parameter $J = 18\ \text{cm}^{-1}$ for the solid solutions with low chromium concentrations and even at the infinite dilution. With an increase in the chromium content in solid solutions and the distribution of an essential number of its atoms over the *B*-sites the antiferromagnetic interactions prevail.

5.3.2.2 Magnetic properties of iron-containing bismuth titanates and niobates

Magnetic behavior of the $Bi_{1.6}Fe_xTi_2O_{7-\delta}$ solid solutions was studied for samples with $0.095 \leq x \leq 0.42$.[77] In the solid solutions with small content of iron its atoms were found to be distributed over the *A* (bismuth) sites. When $x \geq 0.4$ a certain fraction of the iron atoms occupy the *B* (titanium) sites (Fig. 5.26, Table 5.9).

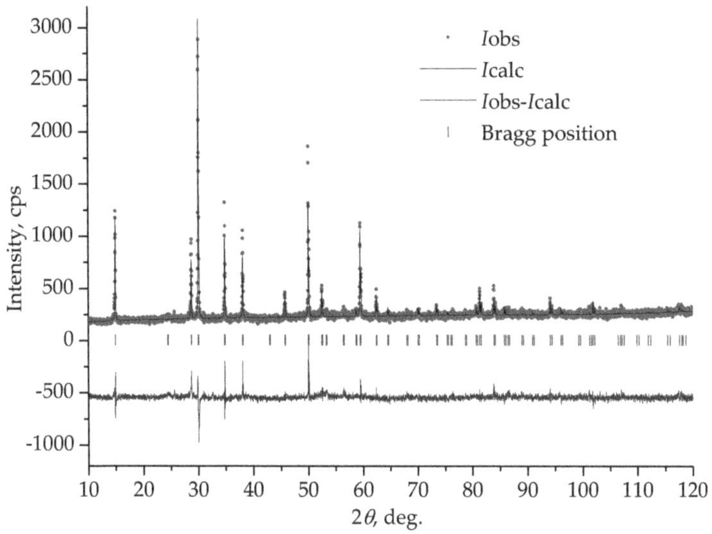

Figure 5.26 Observed, calculated and difference X-ray powder diffraction pattern of $Bi_{1.6}Fe_{0.4}Ti_2O_7$.

Table 5.9 Atomic coordinates (x, y, z), anisotropic thermal parameters (U_{11}, U_{22}, U_{33}), and occupancy in $Bi_{1.6}Fe_{0.4}Ti_2O_{7-\delta}$ by Rietveld refinement (R_p = 6.35%; R_{wp} = 8.29%; χ^2 = 1.75)

Atom	Site	*x*	*y*	*z*	U_{11}, U_{22}	U_{33}	Occ.
Bi/Fe	16c	0	0	0	0.89(1)/ 0.89(4)	0.89(1)/ 0.89(4)	0.8/0.2
Bi/Fe	96*g*	0.0212	0.0254(6)	0.0212(7)	0.88(4)/ 0.27(3)	0.71(1)/ 1.07(3)	0.76/0.15
Ti/Fe	16*d*	1/2	1/2	1/2	1.28(5)/ 1.28(5)	1.28(5)/ 1.28(5)	0.95/0.05
O	48*f*	1/8	1/8	0.428(1)	0.20(1)	0.21(3)	1
O′	32*e*	0.1128	0.1128(4)	0.1128(4)	1	1	0.7

Note: Anisotropic thermal parameters units Å × 100; a = 10.3328(6) Å. $(Bi_{1.52}Fe_{0.3})(Fe_{0.1}Ti_{1.9})O_{6.7}$—model ($R_p$ = 6.35%; R_{wp} = 8.29%; χ^2 = 1.75).

The oxidation state of iron atoms in the bismuth titanate pyrochlores was studied by Mössbauer spectroscopy.[77] Mössbauer spectra for $Bi_{1.6}Fe_{0.1}Ti_2O_{6.4}$ and $Bi_{1.6}Fe_{0.4}Ti_2O_7$ compositions are given in Fig. 5.27. The parameters of the spectra are given in Table 5.10.

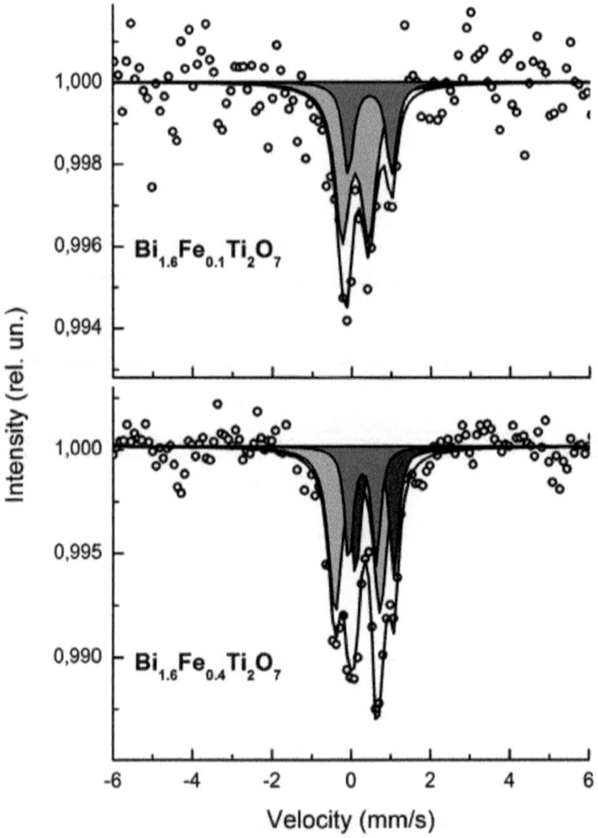

Figure 5.27 Mössbauer spectra of $Bi_{1.6}Fe_{0.1}Ti_2O_{6.4}$ and $Bi_{1.6}Fe_{0.4}Ti_2O_7$ at 298 K.

Both compositions are characterized by the presence of iron in the oxidation state +3.

The spectrum of the $Bi_{1.6}Fe_{0.1}Ti_2O_{6.4}$ is described by two doublets. In $Bi_{1.6}Fe_{0.40}Ti_2O_7$ a greater number of the states of iron(III) seems to be realized, which is manifested by three doublets in the spectrum.

Table 5.10 Parameters of Mössbauer spectra of $Bi_{1.6}Fe_xTi_2O_{7-\delta}$ at 298 K

Composition		Isomeric shift (IS) ($mm \cdot s^{-1}$)	Quadrupole splitting (QS) ($mm \cdot s^{-1}$)	Doublet fraction (%)
$Bi_{1.6}Fe_{0.4}Ti_2O_7$	1	0.156(24)	1.140(53)	51(2) (*A*-site)
	2	0.277(25)	0.689(47)	22(3) (*B*-site)
	3	0.593(20)	0.974(33)	27(2) (*A*-site)
$Bi_{1.6}Fe_{0.1}Ti_2O_{6.4}$	1	0.092(37)	0.676(63)	70(3) (*A*-site)
	2	0.458(46)	1.098(97)	30(4) (*A*-site)

Doublets in the spectra of bismuth titanates and niobates pyrochlores differ significantly both in the isomeric shift (IS) and in the values of quadrupole splitting (QS) not only for various compositions, but for the same composition. Large QS point to strong distortions in the oxygen polyhedron of iron atoms.

In the iron-containing bismuth titanates pyrochlores the iron atoms are distributed mainly over *A* (bismuth) sites at $x < 0.4$, for $Bi_{1.6}Fe_{0.40}Ti_2O_7$ a fraction of iron atoms is distributed also over *B* (titanium) sites. If we accept the formula taking into account the distribution of iron atoms in $Bi_{1.6}Fe_{0.4}Ti_2O_7$ over two cation sites $(Bi_{1.52}Fe_{0.29}\square_{0.19})(Fe_{0.1}Ti_{1.9})O_6O'_{0.67}$ ($\square$ denotes a vacancy), then about 33% of oxygen (O′) vacancies are formed in the A_4O' sublattice. Then an atom in the *A*-sites (Bi or Fe) can be surrounded by eight oxygen atoms (6O + 2O′ axially), or seven, or only six oxygen atoms, from the BO_6 sublattice, which forms a corrugated hexagonal ring round the atom in *A*-site. Near iron atoms one or two oxygen vacancies can be located along the threefold axis instead of the O′ oxygen atoms (Fig. 5.17).

On the basis of the examination of Mössbauer spectra iron-containing bismuth titanates with the pyrochlore-type structure, we can expect several states of iron atoms (two or three) in the bismuth sublattice with different geometry and different degree of distortion of oxygen surrounding and one state of iron atoms in the *B* (titanium) sites. For the composition with low iron content and crystallographic formula $(Bi_{1.6}Fe_{0.1}\square_{0.3})Ti_2O_{6.4}$, it is logical to assume that iron atoms are distributed over the *A* (bismuth) sites, and, taking into account the fraction of these sites remaining unoccupied, this means about 60% of the oxygen O′ vacancies are in the A_4O' sublattice. Therefore, it is unlikely that the atoms in *A*-sites (including iron atoms) are surrounded by eight oxygen atoms (6O + 2O′). Thus, two states of iron, described by two doublets in the spectra can correspond to iron atoms in A_4O' sublattice with different oxygen surrounding and its distortion, and the third doublet for $Bi_{1.6}Fe_{0.4}Ti_2O_7$ can be due to iron atoms, distributed over *B*-sites. According to the results of the study, it can be concluded that there are two types of iron atoms with the 3+ oxidation state in the distorted oxygen octahedra in $Bi_{1.6}Fe_{0.1}Ti_2O_{7-\delta}$ pyrochlore and three types in $Bi_{1.6}Fe_{0.4}Ti_2O_7$. These data also confirm the preferential distribution of iron atoms in the *A* (bismuth) sites and the possibility of getting them in the *B* (titanium) sites in case of an increase in the iron content.

The EPR spectrum of $Bi_{1.6}Fe_{0.4}Ti_2O_7$ powder was measured at room temperature. An anisotropic broad signal with $g_\perp \approx 4.3$ and $g_{||} \approx 2.1$ corresponds to Fe^{3+} in a highly distorted cubic and octahedral oxygen surrounding.

The dependence of $\chi_{Fe}(x)$ and μ_{eff} ($x \to 0$) values for $Bi_{1.6}Fe_x Ti_2O_{7-\delta}$ ($0.095 \le x \le 0.42$) are given in Fig. 5.28 and Table 5.11. The changes in χ^{-1} with the temperature obey Curie–Weiss law with negative Weiss constant (Table 5.11).

The magnetic moment of the solid solution with $x = 0.095$ increases from 5.49 μB (77 K) to 5.84 μB (400 K). For compositions with $x = 0.35$; 0.42, the magnetic moment is noticeably lower, it increases from 4.5 μB to 5.4 μB when the sample is heated, which, along with the type of concentration dependence of magnetic susceptibility, points to antiferromagnetic exchange between iron atoms. Determination of the exchange parameter and the fraction of dimers Fe^{3+}–O–Fe^{3+} was carried out in accordance with the previously considered model for $S_1 = S_2 = 5/2$, $g = 2$. $J = -17\text{cm}^{-1}$ was found in the calculation, which is noticeably less than the obtained values of the exchange parameter in bismuth iron-containing titanates with layered perovskite structure. This magnetic behavior must be determined by a decrease in the angle of exchange in the TiO_6 sublattice accompanied by an increase in the contribution of ferromagnetic exchange channels. The fraction of dimers is increased with increasing iron concentration (Table 5.12).[66]

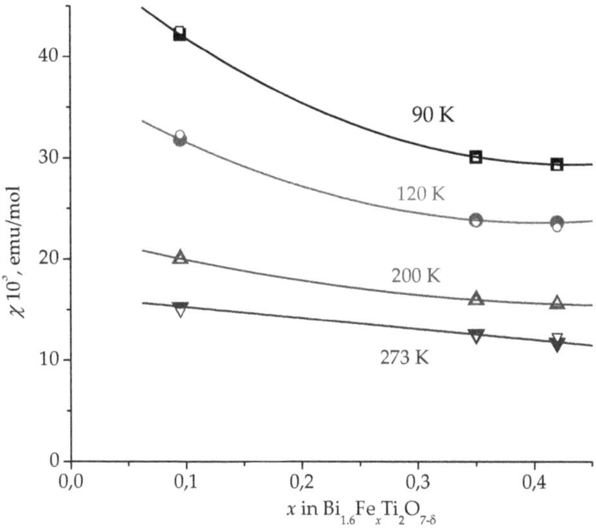

Figure 5.28 Dependence of magnetic susceptibility on the concentration of iron in $Bi_{1.6}Fe_xTi_2O_{7-\delta}$.

Table 5.11 The effective magnetic moment of iron, Weiss constant (θ) and the fraction of dimers in $Bi_{1.6}Fe_xTi_2O_{7-\delta}$

		μ_{eff} (μB)			$a_2 \pm \Delta(a_2)$
x	θ(K)	77 K	180 K	293 K	(J = −17 cm^{-1})
0.095	−16	5.49	5.69	5.81	0.15 ± 0.04
0.35	−51	4.59	4.99	5.29	0.47 ± 0.01
0.42	−40	4.50	4.96	5.13	0.49 ± 0.03

Table 5.12 Effective magnetic moments and Weiss constants for iron-containing bismuth niobates pyrochlores

$n(Bi_2O_3):n(Fe_2O_3):n(Nb_2O_5)$	Formula	μ_{eff} (μB)	θ(K)
0.4400:0.2700:0.2900	$Bi_{1.721}Fe_{1.056}Nb_{1.134}O_7$	4.87	−233
0.4350:0.2850:0.2800	$Bi_{1.711}Fe_{1.121}Nb_{1.101}O_7$	4.91	−278
0.4250:0.2800:0.2950	$Bi_{1.657}Fe_{1.092}Nb_{1.150}O_7$	5.21	−240
0.4200:0.2950:0.2850	$Bi_{1.647}Fe_{1.157}Nb_{1.118}O_7$	5.23	−273
0.4200:0.2150:0.3650	$Bi_{1.576}Fe_{0.8070}Nb_{1.370}O_7$	4.84	−164

The results of the study of magnetic properties of concentrated iron-containing bismuth niobates with the pyrochlore-type structure are reported in ref. 66. The temperature dependence of the inverse magnetic susceptibility of the compounds is linear above 150 K, Weiss constants for a number of samples and their effective magnetic moments are given in the Table 5.12. Bi–Fe–Nb-O′ pyrochlores are paramagnetic with antiferromagnetic exchange at low temperatures and some supraparamagnetic effects defining a decrease in the effective magnetic moments for high-spin Fe^{3+}.

Large negative Weiss constants correlating with iron concentration in the iron-containing bismuth niobates pyrochlores point to significant antiferromagnetic exchange interactions between iron atoms located in *B*-sites. In bismuth niobates, iron atoms are distributed over *B*-sites to a greater extent compared to the bismuth titanate pyrochlores, hence antiferromagnetic interactions are stronger.

In comparison with similar bismuth niobate pyrochlores in Bi–*M*–Nb–O′ systems (*M*—Mn, Ni, Mg, Co), a large "pyrochlore field" existing in the phase diagram of Bi–Fe–Nb–O′ system points to the fact that the pyrochlore structure is tolerant to essentially lower iron concentrations in *A*-sites and therefore the vacancy concentration in *A*-sites will be noticeably higher. These features are explained by the presence of trivalent iron, while in other systems the substituting *M* cations are partially or completely present in the bivalent state.[66]

Thus, for iron-containing bismuth titanates and niobates with both types of structures antiferromagnetic exchange interactions exist in the systems with a high level of *B*-sites substitution.

5.3.2.3 Magnetic properties of manganese-containing bismuth titanates and niobates

Magnetic behavior of the $Bi_{1.6}Mn_xTi_2O_{7-\delta}$ and $Bi_{1.6}Mn_{0.8+x}Nb_{1.6}O_{7-\delta}$ pyrochlore solid solutions were studied.[78,65] In was found, that both systems are characterized by a wide range of manganese concentrations with pyrochlore structure ($0.016 \le x \le 1.23$ and $0 \le x \le 0.8$, respectively).

The results of the structural analysis of manganese-containing bismuth titanate are given in Fig. 5.29 and Table 5.13.

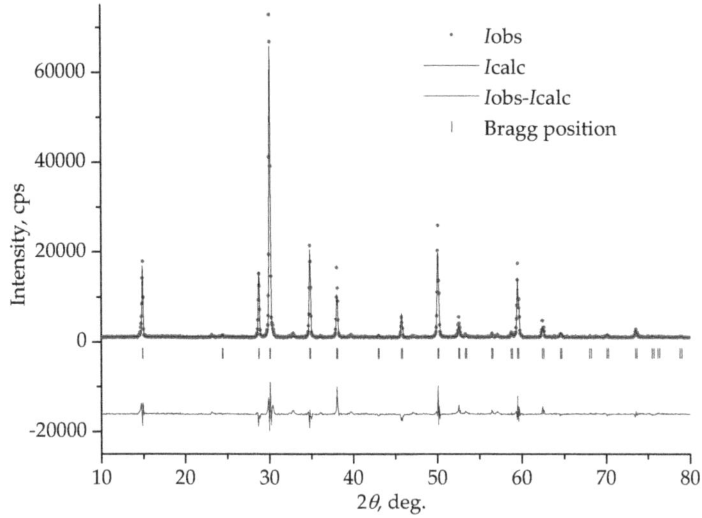

Figure 5.29 Observed, calculated, and difference X-ray diffraction profiles for the $Bi_{1.6}Mn_{0.8}Ti_{1.6}O_{7-\delta}$ pyrochlore at 298 K.

Table 5.13 Atomic coordinates (x, y, z), anisotropic thermal parameters (U_{11}, U_{22}, U_{33}), and occupancy in $Bi_{1.6}Mn_{0.8}Ti_{1.6}O_{6.4}$ by Rietveld refinement (R_p = 5.26%; R_{wp} = 7.43%; χ^2 = 4.48 for $Bi_{1.54}Mn_{0.31}Mn_{0.46}Ti_{1.54}O_{6.16}$)

Atom	Site	x	y	z	U_{11}/U_{22}	U_{33}	Occ.
Bi/Mn	16*c*	0	0	0	0.267(4)/ 0.018(3)	0.267/ 0.018	0.77/ 0.16
Ti/Mn	16*d*	1/2	1/2	1/2	0.153(3)/ 0.092(5)	0.153/ 0.092	0.79/ 0.21
O	48*f*	1/8	1/8	0.434(3)	0.503(1)	2.65(3)	1
O′	8*a*	1/8	1/8	1/8	0.101(7)	0.101	0.21

Note: Anisotropic thermal parameters units Å × 100; a = 10.3002(5) Å.

The oxidation state of manganese was studied by EXAFS spectroscopy. NEXAFS Mn2*p* spectra of manganese-containing bismuth titanate and niobate pyrochlores and of MnO, Mn_2O_3, MnO_2 manganese oxides are shown in Figs. 5.30 and 5.31.

The NEXAFS Mn2*p*-spectra examination (Fig. 5.31) points to the following: (i) the Mn2*p*-spectra of manganese oxides (MnO, Mn_2O_3, MnO_2) obtained in our study correspond to the Mn2*p*-spectra studied earlier by Gilbert and coworkers,[79] (ii) the spectra of bismuth manganese titanate pyrochlores with various manganese content are similar both in shape of the absorption lines and their energies, (iii) the fine structure Mn2*p*-spectra of titanate pyrochlores are identical to those for MnO. For the $Bi_{1.56}Mn_{1.17}Nb_{1.17}O_{7-\delta}$ pyrochlore according to EELS (electron energy-loss spectroscopy) manganese was found to be in +3 oxidation state—Mn^{3+}[80] or in Mn^{2+} and Mn^{3+} oxidation states Mn^{2+} and Mn^{3+} (Fig. 5.31a).[65]

The authors conclude that for $Bi_{1.6}Mn_{1.2}Nb_{1.2}O_{7-\delta}$ pyrochlore manganese atoms in differ oxidation states are distributed over the sites: Mn^{2+} in the *A*(16*d*)-sites, Mn^{3+} in the *B*(16*c*)-sites with the formula $Bi_{1.6}Mn^{2+}{}_{0.4}(Mn^{3+}{}_{0.8}Nb_{1.2})O_7$ and with the occupation (full filling) of all of the cationic and anionic positions.

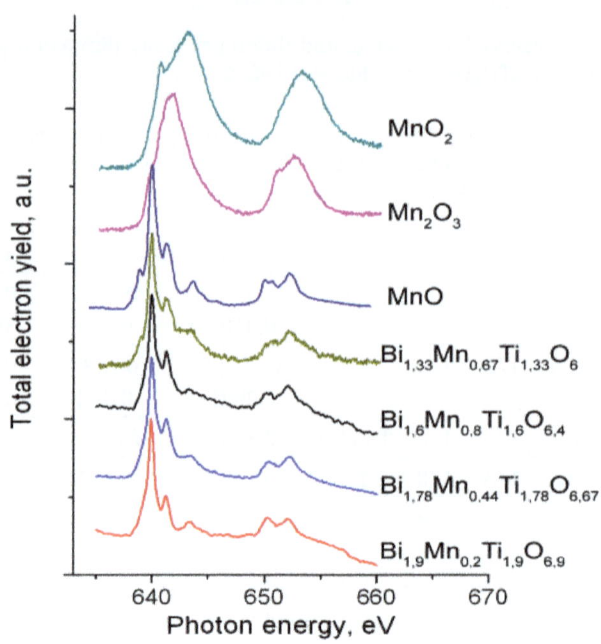

Figure 5.30 NEXAFS Mn2*p* spectra for the $Bi_2Mn_xTi_2O_{7-\delta}$ pyrochlores and manganese oxides.[78]

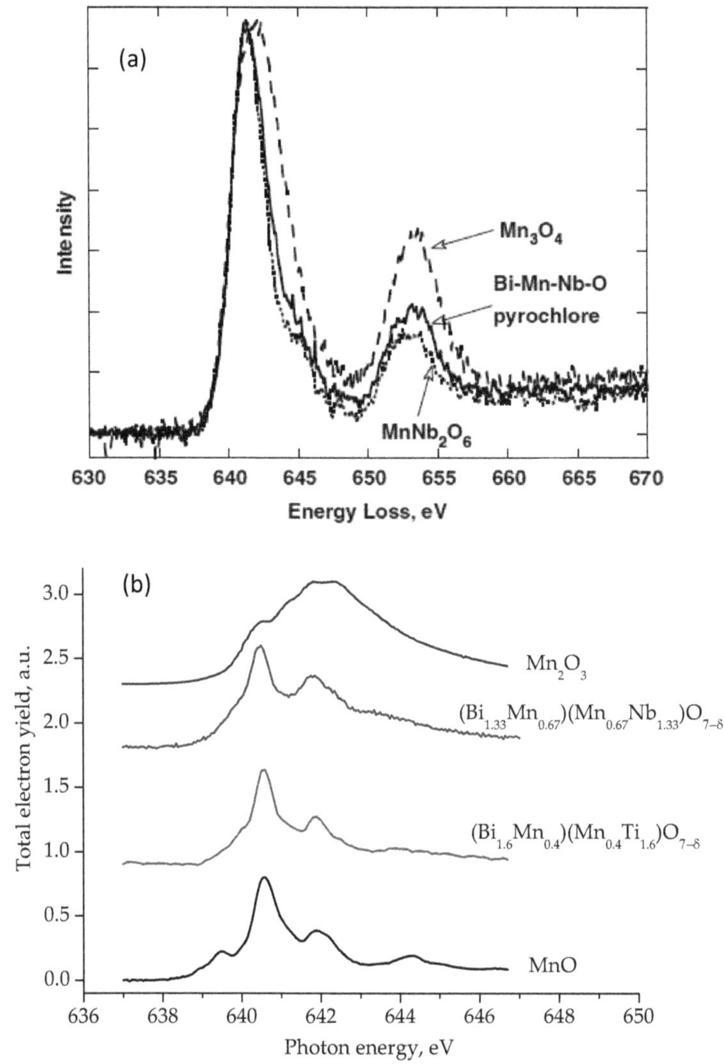

Figure 5.31 $MnL_{2,3}$ EELS spectra of $Bi_{1.6}Mn_{1.2}Nb_2O_{7-\delta}$ pyrochlore, Mn_3O_4, $MnNb_2O_6$[65] (a); NEXAFS Mn2*p* spectra of $(Bi_{1.33}Mn_{0.67})(Mn_{0.67}Nb_{1.33})O_{7-\delta}$ and $(Bi_{1.33}Mn_{0.4})(Mn_{0.4}Ti_{1.33})O_{7-\delta}$ pyrochlores, MnO, Mn_2O_3 (b).

The NEXAFS Mn2*p* spectra of manganese-containing titanate and bismuth niobate with the pyrochlore structure with the same number of manganese atoms—$Bi_{1.33}Mn_{1.33}Nb_{1.33}O_{7-\delta}$ and $Bi_{1.33}Mn_{1.33}Ti_{1.33}O_{7-\delta}$—are given in Fig. 5.31b. In these compositions, manganese is distributed over *A* and *B* cation sites in equal parts, which corresponds to the formula $Bi_{1.33}Mn^{+2}_{0.67}(Mn^{+3}_{0.67}Nb_{1.33})O_{7-\delta}$ with full filling of the sites or $Bi_{1.33}Mn^{+2}_{0.67}(Mn^{+2}_{0.67}Nb_{1.33})O_{6.65}$ with vacancies in oxygen positions (O′). Similar distribution of manganese 2+ is to be expected for bismuth titanate pyrochlore. NEXAFS Mn2*p* spectra for bismuth titanate and niobate with pyrochlore-type structure are slightly different in the shape of the lines. The shape of absorption lines and their energies for manganese-containing bismuth titanate are similar to those for Mn2*p*-spectrum of the bismuth titanate with a lower content of manganese, as well as to the spectrum of MnO, so, manganese is present in them in the form of Mn^{2+}. The spectrum of manganese-containing bismuth niobate pyrochlore to a greater extend corresponds to the superposition of Mn2*p* spectra of Mn^{2+} and Mn^{3+} oxides.

On the basis of the above mentioned (Fig. 5.31), we believe that the manganese oxidation state is mainly Mn^{2+} in bismuth manganese titanates, whereas in similar Bi-Mn-Nb-O pyrochlores it is believed to be 3+.

Magnetic behavior of the solid solutions $Bi_{1.6}Mn_xTi_2O_{7-\delta}$ was studied for samples with $0.016 \leq x \leq 1.23$. The dependence of $\chi^{-1}(T)$ for manganese-containing bismuth titanate pyrochlores, $Bi_{1.6}Mn_xTi_2O_{7-\delta}$ obeys Curie–Weiss law with Weiss constant decreasing as the manganese content increases to $\theta = -68$ K (at $x = 1.23$). χ_{Mn} increases with dilution (Fig. 5.32a). The effective magnetic moment for compounds with low manganese concentration ($x < 0.1$ and at the infinite dilution ($x \rightarrow 0$) $6.6 \leq \mu_{\mathrm{eff}} \leq 7.2$ exceeds 5.92 μB (the spin-only value of μ_{eff} for Mn^{2+}, (d^5, $^6A_1$)) (Fig. 5.32b).

For the samples with $x \geq 0.24$ the magnetic moments are significantly lower and slightly increase as the temperature increases from 3.5 μB (77 K) to 3.8 μB ($x = 0.8$) and to 4.5 μB ($x = 1.23$) at 400 K. An increase in the effective magnetic moments along with a large negative Weiss constant points to antiferromagnetic interaction between manganese atoms in magnetically concentrated solutions.

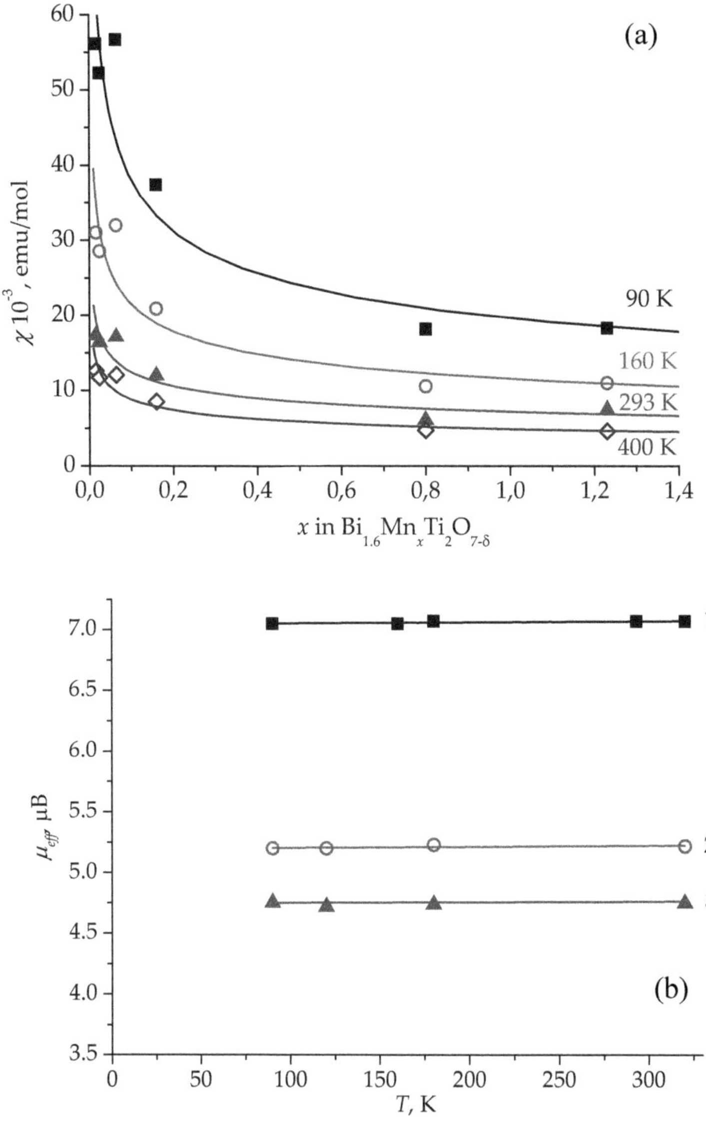

Figure 5.32 Concentration dependence of magnetic susceptibility of $Bi_{1.6}Mn_xTi_2O_{7-\delta}$ (a) and temperature dependence of the effective magnetic moment of manganese in $Bi_{1.6}Mn_xTi_2O_{7-\delta}$ (1), $Bi_{1.6}Mg_{0.8(1-x)}Mn_{0.8x}Nb_{1.6}O_{7-\delta}$ (2), $Bi_{1.6}Zn_{0.8(1-x)}Mn_{0.8x}Nb_{1.6}O_{7-\delta}$ (3) at the infinite dilution ($x \to 0$) (b).

Concentration dependence of $\chi_{Mn}(x)$ is typical for the dilution of antiferromagnets. However, the extrapolation of $\chi_{Mn}(x)$ to the infinite dilution ($x \to 0$) results in the effective magnetic moment μ_{eff} 6.8–7.2 μB and it does not depend on temperature (Fig. 5.32b). This fact could not be explained from the standpoint of only single manganese atoms. It must be assumed that in strongly diluted solid solutions some ferromagnetically bonded aggregates, possibly dimers are formed, which with increasing concentration of manganese in the compound become aggregated and interact antiferromagnetically. A ferromagnetic component in the exchange interaction between atoms with $^6A_{1g}$ ground state can be explained only by a significant deviation of the exchange angle from 180°. If we assume that at the infinite dilution manganese atoms are aggregated and form dimers ($S_1 = S_2 = 5/2$), the fraction of such dimers and the exchange parameter for compounds with a low manganese content, including the infinite dilution ($x \to 0$), essentially differ from these characteristics for compounds with a high concentration of manganese (Table 5.14, Fig. 5.32).

Table 5.14 Magnetic susceptibility and effective magnetic moment of $Bi_{1.6-0.5x}Mn_xTi_{2-0.5x}O_{7-\delta}$

T, K	$\chi(Mn)(x \to 0) \cdot 10^3$ (emu·mol^{-1})	μ_{eff} (μB)			
		$x \to 0$	$x = 0.016$	$x = 0.16$	$x = 0.80$
90	64.2	6.8	6.4	5.3	3.7
160	38.3	7.0	6.4	5.3	3.7
180	35.0	7.1	6.4	5.4	3.8
293	21.5	7.1	6.5	5.5	3.9
320	20.3	7.2	6.5	5.5	3.9

Manganese atoms in $Bi_{1.6-0.5x}Mn_xTi_{2-0.5x}O_{7-\delta}$ with $x > 0.4$ are distributed over both cationic sites: A (bismuth) and B (titanium). An increase in the fraction of manganese atoms in the BO_6 sublattice, in which the TiO_6 octahedra are connected by vertices, provides the antiferromagnetic interaction between Mn^{2+}–O–Mn^{2+}. In the compositions with the concentration of manganese $x < 0.4$, its atoms are distributed over the bismuth sites in the Bi_4O' sublattice. Ferromagnetic exchange can occur between neighboring atoms, which are located in different cationic positions. The exchange interactions in Bi_4O' sublattice of pyrochlore-type structure are not possible geometrically. The results of calculations based on the assumption about of the formation of ferromagnetic dimers in $Bi_{1.6}Mn_xTi_2O_{7-\delta}$ and optimal parameters of exchange are given in Table 5.15. The fractions of ferromagnetic dimers in the infinitely diluted solution and in $Bi_{1.6}Mn_{0.016}Ti_2O_{7-\delta}$ are the same. Ferromagnetic interaction most probably is preserved in other compositions, with larger manganese concentration. The fraction of antiferromagnetic aggregates (dimers—in the first approximation) for manganese concentrated samples was determined (Table 5.15).

Table 5.15 The fraction of dimers of $Bi_{1.6}Mn_xTi_2O_{7-\delta}$ ($x \leq 0.4$) and $Bi_{1.6-0.5x}Mn_xTi_{2-0.5x}O_{7-\delta}$ ($x > 0.4$)

x	$a_2^{(fm)} \pm \Delta(a_2)$ $J = 30$ cm^{-1}	$a_2^{(afm)} \pm \Delta(a_2)$ $J = -40$ cm^{-1}
$x \to 0$	0.23 ± 0.07	—
0.016	0.20 ± 0.05	—
0.024	0.20 ± 0.05	0.15 ± 0.02
0.16	0.20 ± 0.05	0.50 ± 0.05
0.80	0.20 ± 0.05	0.80 ± 0.10
1.23	0.20 ± 0.05	0.80 ± 0.04

The study of magnetization from 4 to 300 K in the fields of 0–5 T for $Bi_{1.6}Ti_2Mn_{0.1}O_{7-\delta}$ was carried out. The data obtained from these measurements agree well with the data obtained using Faraday method in the range of 77–400 K. The ferromagnetic contribution at low temperatures is confirmed by the nonlinear dependence of magnetization of the sample on the field at 4 K (Fig. 5.33). When the sample is heated to room temperature, the ferromagnetic contribution almost disappears. Antiferromagnetic interaction extends all over the lattice and becomes cooperative most probably in the samples with high manganese content.

Thus, ferromagnetically coupled aggregates to some extent arising between neighboring atoms located in *A* (bismuth) and *B* (titanium) sites may determine an increase in the effective magnetic moment in comparison with the spin-only value for manganese-containing bismuth titanates with pyrochlore-type structure.

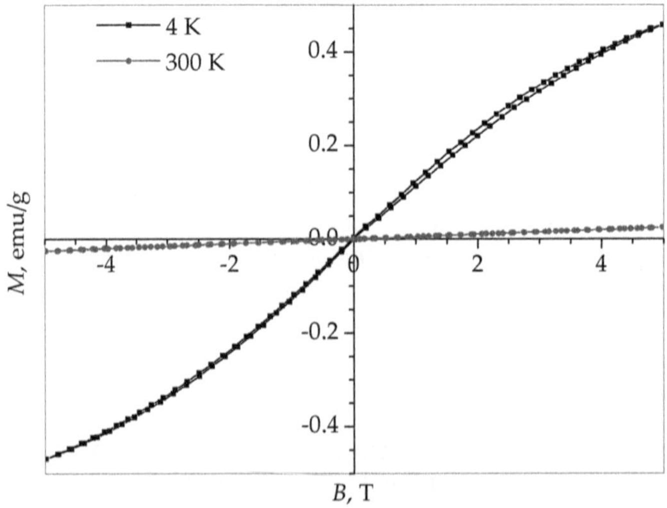

Figure 5.33 The dependence of specific magnetization of $Bi_{1.6}Ti_2Mn_{0.1}O_{7-\delta}$ on the field at 4 K and 300 K.

When the manganese concentration increases ($x \geq 0.16$), magnetic susceptibility decreases owing to arising antiferromagnetic exchange interactions. In manganese-containing bismuth titanate pyrochlores antiferromagnetically coupled dimers appear ($J = -40\ cm^{-1}$; a_2 increases from 0.5 to 0.8 with increasing x in $Bi_{1.6-0.5x}Mn_xTi_{2-0.5x}O_{7-\delta}$).

5.3.2.4 Magnetic properties of diluted manganese-containing bismuth niobates

Magnetic properties of diluted manganese-containing bismuth niobates with the pyrochlore-type structure were studied $Bi_2Mg_{1-x}Mn_xNb_2O_{7+\delta}$ (FU—$Bi_{1.6}Mg_{0.8-0.8x}Mn_{0.8x}Nb_{1.6}O_{7-\delta}$) and $Bi_2Zn_{1-x}Mn_xNb_2O_{7+\delta}$ (FU—$Bi_{1.6}Mg_{0.8-0.8x}Mn_{0.8x}Nb_{1.6}O_{7-\delta}$) with x = 0.01–0.2. Containing only manganese as a substituent, bismuth niobates with pyrochlore structure are formed only for compositions with a substantial substituent fraction (≥30% replacement of cationic sites).[65]

The results of a recent study of manganese concentrated bismuth niobate pyrochlores are given in Table 5.16. All compositions are paramagnetic and show weak antiferromagnetic interactions. The authors explain a lower effective magnetic moment compared to the spin-only values by a possible coexistence of Mn^{2+} and Mn^{3+} occupying different cationic *A*- and *B*-sites. The calculated magnetic moment estimated from Mn^{2+} and Mn^{3+} ratios based on the structural analysis data using the neutron diffraction pattern are given in Table 5.16. From our point of view the main cause of the magnetic moment reduction in these compositions is a strong antiferromagnetic exchange. In $Mn_2Sb_2O_7$ pyrochlore[81] similar values of Weiss constant (−33 K) were observed, Mn^{2+} occupying *A*-sites shows a lower magnetic moment (5.76 μB) compared to spin-only value for Mn^{2+}(d^5, 5.92 μB), which, however, is much higher than in the case of manganese-containing bismuth niobate pyrochlores.

Table 5.16 Effective magnetic moments and Weiss constants of manganese-containing bismuth niobates

$n(Bi_2O_3)$:$n(Mn_2O_3)$:$n(Nb_2O_5)$	Formula	μ_{eff} (μB)	θ (K)	μ_{calc} (μB)
0.3600:0.3200:0.3200	$Bi_{1.44}Mn^{2+}{}_{0.56}Mn^{3+}{}_{0.72}Nb_{1.28}O_7$	4.94	−34	5.37
0.4000:0.3000:0.3000	$Bi_{1.60}Mn^{2+}{}_{0.40}Mn^{3+}{}_{0.8}Nb_{1.2}O_7$	4.95	−35	5.26

We studied two types of solid solutions: $Bi_{1.6}Mg_{0.8(1-x)}Mn_{0.8x}Nb_{1.6}O_{7-\delta}$ and $Bi_{1.6}Zn_{0.8-x}Mn_xNb_{1.6}O_{7-\delta}$ as magnetically diluted systems. The magnetic susceptibility of both types of the solid solutions is almost independent on dilution (Figs. 5.34 and 5.35). The effective magnetic moment of $Bi_{1.6}Mg_{0.8(1-x)}Mn_{0.8x}Nb_{1.6}O_{7-\delta}$ compounds resulting from extrapolation of χ_{Mn} with $x \leq 0.08$ is 5.3 μB (5.2 μB at $x \to 0$). It does not depend on concentration and slightly depends on temperature (Fig. 5.32b).

The experimental effective magnetic moment in the manganese and magnesium-containing solid solutions of bismuth niobate with pyrochlore-type structure is decreased in comparison with the spin-only effective magnetic moment and correlates with the value obtained in the assumption of the presence of Mn^{2+} and Mn^{3+} in the solid solutions with 0.41 and 0.59 mole fractions of manganese, respectively. In manganese-containing bismuth niobates with pyrochlore structure, manganese could be in the form of Mn^{2+} and Mn^3, respectively, as x is 0.33–0.40 and 0.67–0.60 (Table 5.16).

In manganese doped zinc-containing bismuth niobates with pyrochlore-type structure, $Bi_{1.6}Zn_{0.8(1-x)}Mn_{0.8x}Nb_{1.6}O_{7-\delta}$, the effective magnetic moment almost does not depend on temperature and concentration of manganese with a low manganese content ($x \leq 0.2$). Its value obtained by extrapolating χ ($x \to 0$) 4.75 μB, is significantly lower than the spin-only value for Mn^{2+} (5.92 μB), lower than for Mn^{3+} (4.90 μB) and higher than for Mn^{4+} (3.87 μB) (Table 5.17, Fig. 5.35).

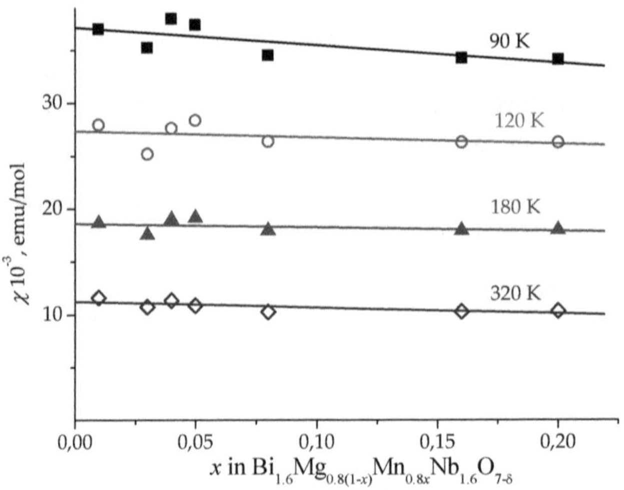

Figure 5.34 Concentration and temperature dependences of paramagnetic component of magnetic susceptibility in $Bi_{1.6}Mg_{0.8(1-x)}Mn_{0.8x}Nb_{1.6}O_{7-\delta}$.

Table 5.17 Effective magnetic moment of manganese in $Bi_{1.6}Zn_{0.8(1-x)}Mn_{0.8x}Nb_{1.6}O_{7-\delta}$ (at extrapolation of χ_{Mn} to the infinite dilution)

	μ_{eff} (μB)	
T (K)	$Bi_{1.6}Zn_{0.8(1-x)}Mn_{0.8x}Nb_{1.6}O_{7-\delta}$	$Bi_{1.6}Mg_{0.8(1-x)}Mn_{0.8x}Nb_{1.6}O_{7-\delta}$
90	4.74	5.22
120	4.75	5.22
180	4.76	5.23
320	4.76	5.24

Such a magnetic behavior can be explained by the coexistence of manganese atoms in different oxidation states: Mn^{2+} and Mn^{4+} (0.43: 0.57) or Mn^{3+} and Mn^{4+} in the ratio 0.85:0.15. It should be noted that magnesium and zinc as dominant substituents have a preference for occupation of different cationic (Mg in *A*, Zn in *B*) sites and thus influences the distribution and oxidation states of manganese atoms in substituted bismuth niobates with pyrochlore-type structure, which is manifested by the magnetic characteristics of the corresponding solid solutions.

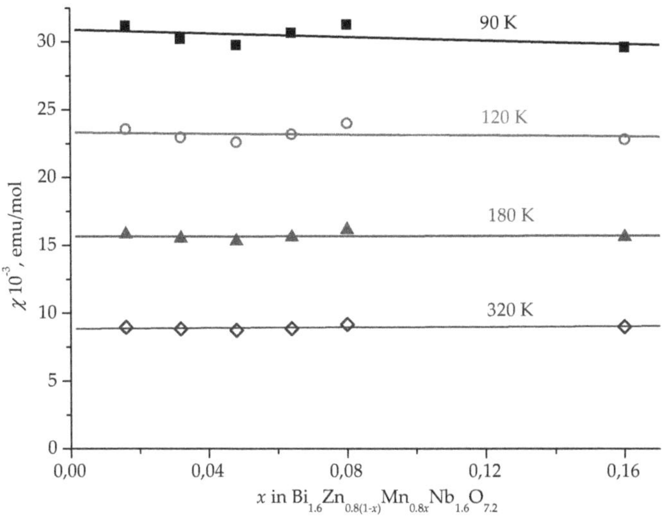

Figure 5.35 Concentration dependence of paramagnetic component of magnetic susceptibility for $Bi_{1.6}Zn_{0.8(1-x)}Mn_{0.8x}Nb_{1.6}O_{7-\delta}$.

5.4 Conclusions

Based on the structural, spectroscopic and magnetochemical study of doped bismuth titanates and niobates containing 3*d*-elements—chromium, iron, manganese, the following trends are revealed.

In the diluted solid solutions of all $Bi_4M_xTi_{3-x}O_{12-\delta}$ and $Bi_5Ti_{3-x}M_xO_{15-\delta}$ (M = Cr, Mn, Fe) under study—bismuth titanates and bismuth niobates with layered perovskite-type structure, paramagnetic atoms are present both in the form of single atoms and in a partially aggregated state. Given rather large concentrations of a paramagnetic atom ($x > 0.1$), antiferromagnetic exchange interactions are found between paramagnetic atoms that are located in octahedral (B) sites in perovskite-like layers. In these layers there is a superexchange via the channels $d_{xz}||p_x||d_{xz}$ and $d_{yz}||p_y||d_{yz}$ (d_π–p_π) between the neighboring paramagnetic atoms at an angle of 180°. In chromium-containing bismuth titanates the interaction is so strong that even in the infinitely diluted solid solutions about 30 mol% of chromium atoms are still bonded into dimers, but in bismuth niobate solid solutions chromium atoms are present in the form of single atoms as well as in iron-containing bismuth titanates. In manganese-containing bismuth titanates and niobates with layered perovskite-like structure, there are two types of exchange interactions between manganese atoms—ferromagnetic superexchange, which results from the presence of manganese in different oxidation states—Mn^{4+} and Mn^{3+} and antiferromagnetic superexchange interaction. The difference in the magnetic behavior of diluted solid solutions appears to point to different trends in the distribution of chromium and iron atoms in the B sites. There are the tendencies of aggregation for chromium and manganese atoms in bismuth titanate solid solutions, iron and manganese atoms in bismuth niobate solid solutions, and the preferential existence in the form of single atoms for iron in bismuth titanate, chromium atoms in bismuth niobate and an aggregation of manganese chromium atoms in bismuth titanate and iron atoms in bismuth niobate.

Thus, in doped bismuth titanates and niobates with a layered perovskite-type structure, at significant concentrations of paramagnetic, an antiferromagnetic exchange is manifested due to the overlapping of orbitals of metal and oxygen atoms at an angle of 180° as the main tendency. When the dopant concentration decreases up to infinite dilution, individual features of paramagnetic atoms begin to manifest themselves.

The magnetic behavior of solid solutions based on bismuth titanate and bismuth niobate with pyrochlore-type structure is determined by the distribution of paramagnetic atoms over two cationic sites of the structure. In diluted $Bi_{1.6}M_xTi(Nb)_2O_{7-\delta}$ ($x < 0.3$) solid solutions, the atoms of paramagnetic elements are distributed preferentially over bismuth (A) sites. The arrangement of the 3d-elements in the A sites does not geometrically favor the overlapping of the orbitals necessary for superexchange interactions. Therefore, in the $Bi_{1.6}Cr_xTi_2O_{7-\delta}$ solid solutions, characterized by low region of the chromium concentration, where stable pyrochlore phase exists and chromium atoms are distributed over A sites, there is no exchange interaction between chromium atoms, as well as for $Bi_{1.6}Fe_x Ti_2O_{7-\delta}$ with $x \leq 0.3$. In the iron-containing solid solutions at $x > 0.3$ the distribution of iron atoms in the titanium sites becomes detectable and the exchange interactions and antiferromagnetic dimers with $J = -18$ cm^{-1} appear. The fraction of the dimers in $Bi_{1.6}Fe_xTi_2O_{7-\delta}$ as in $Bi_4Ti_{3-x}Fe_xO_{12-\delta}$ increases with the iron concentration increasing, that is, there is a direct dependence of the antiferromagnetic exchange emergence on the content of the paramagnetic atom in the TiO_6 sublattice, which provides the conditions of spin-spin interaction. In manganese-containing $Bi_{1.6}Mn_xTi_2O_{7-\delta}$, bismuth titanate is characterized by a large concentration region of substitution ($0.016 \leq x_{Mn} \leq 1.23$), paramagnetic atoms can be distributed over both cationic sites (A and B) of the pyrochlore structure and the effect of coexistence of ferro- and antiferromagnetic superexchange is found. The formation of ferromagnetic dimers with $J = 30$ cm^{-1} ($a_2 \approx 0.2$) is the reason for an increased effective magnetic moment in the diluted solid solutions compared to the spin-only value. Ferromagnetic superexchange may exist between neighboring atoms located in A and B sites. As the content of manganese increases ($x \geq 0.16$), the effective magnetic moment decreases owing to antiferromagnetic exchange ($J = -40$ cm^{-1}).

In the substituted chromium, manganese, iron-containing bismuth niobates pyrochlores antiferromagnetic exchange interactions appear in the concentrated solid solutions when paramagnetic atoms are distributed over B sites. At low concentrations, iron atoms are present in the form of single Fe^{3+} atoms, manganese atoms are in the form of single Mn^{2+}, Mn^{3+} atoms and do not show an appreciable tendency to aggregation. For chromium and magnesium-containing bismuth niobate solid solutions ferromagnetic exchange is likely ($J = 18$ cm^{-1}).

Thus, as a result of the magnetochemical study of doped bismuth titanates and niobates pyrochlores, it was shown that the exchange interactions along with the nature, the oxidation state of the paramagnetic atom are determined by their distribution over the cationic sites. Antiferromagnetic exchange is due to the distribution of paramagnetic atoms over the *B*-sites of the pyrochlore and layered perovskite types structure, whereas the ferromagnetic exchange may be mostly the result of paramagnetic atoms location in *A* and *B* sites of pyrochlore structure, and also to some extent owing to the distortions of the angles in perovskite-like structures. It should be assumed (expect) that the study of the magnetic properties of a wide range of concentrations of the paramagnetic element in a compound with a certain structure provides detailed and relevant information about interatomic interactions and the electronic structure of these compounds. Thus a detailed study of the electron structure of such complicated oxide systems gives us a clue to understanding the differences between their performance as materials and a rout to selecting optimal compositions for particular purposes.

References

1. Ragavan AJ, Adams Dean V (2011) Estimating free energies of formation of titanate ($M_2Ti_2O_7$) and zirconate ($M_2Ti_2O_7$) pyrochlore phases of trivalent lanthanides and actinides, *ISRN Mater Sci*, **Article ID 680785**, 1–8.
2. Yao WF, Wang H, Xu XH, Zhou JT, Yang XN, Zhang Y, Shang SX (2004) Photocatalytic property of bismuth titanate $Bi_2Ti_2O_7$, *Appl Catal A Gen*, **259**, 29–33.
3. Kidchob T, Malfatti L, Maongiu D, Enzo S, Innocenzi P (2010) Sol-gel processing of $Bi_2Ti_2O_7$ and $Bi_2Ti_4O_{11}$ films with photocatalytic activity, *J Am Ceram Soc*, **93**(9), 2897–2902.
4. Zhou H, Park T-J (2006) Synthesis, characterization, and photocatalytic properties of pyrochlore $Bi_2Ti_2O_7$ nanotubes, *J Mater Res*, **21**(11), 2941–2947.
5. Subramanian MA, Toby BH, Ramirez AP, Marshall WJ, Sleight AW, Kwei GH (1996) Colossal magnetoresistance without Mn^{3+}/Mn^{4+} double exchange in the stoichiometric pyrochlore $Tl_2Mn_2O_7$, *Science*, **273**, 81–84.

6. Wang X, Wang H, Yao X (1997) Structure, phase transformation and dielectric properties of pyrochlores containing bismuth, *Am Ceram Soc*, **80**(10), 2745–2748.

7. Hong YP, Ha S, Lee HY, Lee YCh, Ko KH, Kim D-W, Hong HB, Hong KS (2002) Voltage tunable dielectric properties of sputtered Bi_2O_3-ZnO-Nb_2O_5 pyrochlore thin films, *Thin Solid Films*, **419**(7), 188–193.

8. Ren W, Trolien McKinsty S, Randall C, Shrout T (2001) Bismuth zinc niobate pyrochlore dielectric thin films for capactive applications, *J Appl Phys*, **89**(1), 767–774.

9. David P, Clive A, Thomas R (1996) Investigation of the dielectric properties of bismuth pyrochlores, *Solid State Commun*, **100**(7), 529–534.

10. Kamba S, Porokhonskyy V, Pashkin A, Bovtun V, Petzelt J, Nino JC, Trolier-McKinstry S, Lanagan MT, Randall CA (2002) Anomalous broad dielectric relaxation in $Bi_{1.5}Zn_{1.0}Nb_{1.5}O_7$ pyrochlore, *Phys Rev B*, **66**, 054106.

11. Nguyen B, Liu Y, Withers RL (2007) The local crystal chemistry and dielectric properties of the cubic pyrochlore phase in the Bi_2O_3-M^{2+}O-Nb_2O_5 (M^{2+} = Ni^{2+} and Mg^{2+}) systems, *J Solid State Chem*, **180**(2), 549–557.

12. Moreno KJ, Fuentes AF, Garcia-Bariocanal J, Leon C, Santamaria J (2006) Mechanochemical synthesis and ionic conductivity in the $Gd_2(Sn_{1-y}Zr_y)_2O_7$ ($0 \le y \le 1$) solid solution, *J Solid State Chem*, **179**, 323–330.

13. Aurivillius B (1949) Mixed bismuth oxides with layer lattices. Structure of $Bi_4Ti_3O_{12}$, *Arkiv for Kemi*, **1**(58), 499–512.

14. Lomanova NA, Ugolkov VL, Panchuk VV, Semenov VG (2017) Formation and thermal behavior of Aurivillius phases $A_{m-1}Bi_2Fe_{m-3}Ti_3O_{3m+3-\delta}$ (A = Bi, Sr), *Russ J Gen Chem*, **87**(3), 365–372.

15. Lomanova NA, Gusarov VV (2011) Phase states in the $Bi_4Ti_3O_{12}$-$BiFeO_3$ section in the Bi_2O_3-TiO_2-Fe_2O_3 system, *Russ J Inorg Chem*, **56**(4), 616–620.

16. Lomanova NA, Gusarov VV (2011) Electrical properties of perovskite-like compounds in the Bi_2O_3-Fe_2O_3-TiO_2 system, *Inorg Mater*, **47**(4), 420–425.

17. Lomanova NA, Morozov MI, Ugolkov VL, Gusarov VV (2006) Properties of Aurivillius phases in the $Bi_4Ti_3O_{12}$-$BiFeO_3$ system, *Inorg Mater*, **42**(2), 189–195.

18. Kim YI, Jeon MK (2004) Combined structural refinement of $Bi_4Ti_3O_{12}$ using X-ray and neutron powder diffraction data, *Mater Lett*, **58**(40890), 1889–1893.

19. Dorrian JF, Newnham RE, Smith DK, Kay MI (1971) Crystal structure of $Bi_4Ti_3O_{12}$, *Ferroelectrics*, **3**(1), 17–27.

20. Rae AD, Thompson JG, Withers RL, Willis AC (1990) Structure refinement of commensurately modulated bismuth titanate $Bi_4Ti_3O_{12}$, *Acta Cryst* B, **46**, 474–487.

21. Hervoches CH, Lightfoot P (1999) A variable-temperature powder neutron diffraction study of ferroelectric $Bi_4Ti_3O_{12}$, *Chem Mater*, **11**(11), 3359–3364.

22. Cummins SE, Cross LE (1968) Electrical and optical properties of ferroelectric $Bi_4Ti_3O_{12}$ single crystals, *J Appl Phys*, **39**(5), 2268–2274.

23. Jeon MK, Woo SI, Kim Y-I, Nahm S-H (2004) Structure analysis of ferroelectric $Bi_4Ti_3O_{12}$ by using x-ray powder diffraction, *J Korean Phys Soc*, **45**(5), 1240–1243.

24. Chakraborty KR, Achary SN, Patwe SJ, Krichna PSR, Shinde AB, Tyagi AK (2007) Low temperature neutron diffraction studies on $Bi_4Ti_3O_{12}$, *Ceram Int*, **33**(4), 601–604.

25. Kan YM, Wang PL, Li YB, Cheng YB, Yan DS (2002) Low-temperature sintering of $Bi_4Ti_3O_{12}$ derived from a co-precipitation method, *Mater Lett*, **56**(6), 910–914.

26. Liu WL, Xia HR, Han X, Wang XQ (2005) Synthesis and structure of bismuth titanate nanopowders prepared by metalorganic decomposition method, *J Mater Sci*, **40**, 1827–1829.

27. Lu C-D, Chang L-S, Lu Y-F, Lu F-H (2009) The growth of interfacial compounds between titanium dioxide and bismuth oxide, *Ceram Int*, **35**(7), 2699–2704.

28. Iwata M, Ando K, Maeda M, Ishibashi Y (2013) Structural phase transition and symmetry of parent phase in $Bi_4Ti_3O_{12}$, *J Phys Soc Jpn*, **82**(2), 025001.

29. Knyazev AV, Krasheninnikova OV, Smirnova NN, Shushunov AN, Syrov EV, Blokhina AG (2015) Thermodynamic properties and X-ray diffraction of $Bi_4Ti_3O_{12}$, *J Therm Anal Calorim*, **122**(2), 747–754.

30. Stojanovic BD, Paiva-Santos CO, Cilense M, Jovalekic C, Lazarevic ZZ (2008) Structure study of $Bi_4Ti_3O_{12}$ produced via mechano-chemically assisted synthesis, *Mater Res Bull*, **43**(7), 1743–1753.

31. Jeon MK, Kim YI, Nahm SH, Sohn JM, Jung CH, Woo SI (2007) Crystal structure and spontaneous polarization of $Bi_{4-x}Nd_xTi_3O_{12}$ studied by using neutron powder diffraction data, *J Phys D: Appl Phys*, **40**(15), 4647–4652.

32. Oh SJ, Shin Y, Tran TT, Lee DW, Yoon A, Halasyamani PS, Ok KM (2012) Structure-property relationships in solid solutions of noncentrosymmetric aurivillius phases, $Bi_{4-x}La_xTi_3O_{12}$ (x = 0–0.75), *Inorg Chem*, **51**(19), 10402–10407.

33. Ye W, Lu Ch, You P, Liang K, Zhou Y (2013) Determination of crystal symmetry for $Bi_4Ti_3O_{12}$-based ferroelectrics by using electron diffraction, *J Appl Crystallogr*, **46**(3), 798–800.

34. Gopalakrishnan J, Ramanan A, Rao CNR, Jefferson DA, Smith DJ (1984) A homologous series of recurrent intergrowth structures of the type $Bi_4A_{m+n-2}B_{m+n}O_{3(m+n)+6}$ formed by oxides of the Aurivillius family, *J Solid State Chem*, **55**, 101–105.

35. Takenaka T, Komura K, Sakata K (1996) Possibility of new mixed bismuth layer-structured ferroelectrics, *Jpn J Appl Phys, Part 1*, **35**(9B), 5080–5083.

36. Tahara S, Shimada A, Kumada N, Sugahara Y (2007) Characterization of $Bi_5Nb_3O_{15}$ by refinement of neutron diffraction pattern, acid treatment and reaction of the acid-treated product with *n*-alkylamines, *J Solid State Chem*, **180**(9), 2517–2524.

37. Srinivas A, Suryanarayana SV, Kumar GS, Kumar MM (1999) Magnetoelectric measurements of $Bi_5FeTi_3O_{15}$ and $Bi_6Fe_2Ti_3O_{18}$, *J Phys: Condens Matter*, **11**(16), 3335–3340.

38. Singh RS, Bhimasankaram T, Kumar GS, Suryanarayana SV (1994) Dielectric and magnetoelectric properties of $Bi_5FeTi_3O_{15}$, *Solid State Comm*, **91**(7), 567–569.

39. Mao XV, Wang W, Chen XB (2008) Electrical and magnetic properties of $Bi_5FeTi_3O_{15}$ compound prepared by inserting $BiFeO_3$ into $Bi_4Ti_3O_{12}$, *Solid State Comm*, **147**(5–6), 186–189.

40. Wu M, Tian Zh, Yuan S, Huang Zh (2012) Magnetic and optical properties of the Aurivillius phase $Bi_5Ti_3FeO_{15}$, *Mater Lett*, **68**, 190–192.

41. Giddings AT, Stennett MC, Reid DP, McCabe EE, Greaves C, Hyatt NC (2011) Synthesis, structure and characterization of the n = 4 Aurivillius phase $Bi_5Ti_3CrO_{15}$, *J Solid State Chem*, **184**(2), 252–263.

42. Kalinnikov VT, Rakitin YuV (eds) (1980) *Introduction to Magnetochemistry. Method of Static Magnetic Susceptibility in Chemistry*, Nauka, Moscow [in Russian].

43. Rakitin YuV, Kalinnikov VT (eds) (1994) *Modern Magnetochemistry*, Nauka, St. Petersburg [in Russian].

44. Chezhina NV (2006) *Magnetochemistry. in Physical Methods of Research in Inorganic Chemistry,* ed, Nikolsky AB". Academy, Moscow [in Russian].

45. Koroleva MS (2014) Structure and physico-chemical properties of doped bismuth titanates $Bi_{1.6}M_xTi_2O_{7-\delta}$ and $Bi_4Ti_{3-x}M_xO_{12-\delta}$ (M–Cr, Fe): Diss. kand. chem. Sciences. Syktyvkar [in Russian].

46. Zhuk NA, Piir IV, Pimenov AL, Chezhina NV (2007) Structure and magnetic and electric properties of bismuth niobates doped with d elements: IV. Magnetic behavior of chromium-containing solid solutions with the structure of bismuth niobate, $Bi_5Nb_3O_{15}$, *Russ J Gen Chem*, **77**(6), 990–996.

47. Chezhina NV, Piir IV, Zhuk NA (2005) State of chromium in solid solutions of complex bismuth niobates $Bi_2BaCr_xNb_{2-x}O_\delta$ ($0.01 < x < 0.3$), *Russ J Gen Chem*, **75**(1), 21–24.

48. Chezhina NV, Korolev DA (2012) State of atoms and interatomic interactions in complex perovskite-like oxides: XXX. Influence of the nature of diamagnetic substituents on the dynamics of clustering in lanthanum gallate doped with strontium, chromium, and magnesium, *Russ J Gen Chem*, **82**(3), 347–353.

49. Chezhina NV, Korolev DA, Fedorova AV, Zhuk NA, Butin VA, Lutoev VP, Makeev BA, Shevchuk SS, Nizovtsev AN (2017) Structure, magnetic, and electrical properties of bismuth niobates doped with *d*-elements: XV.1 Exchange interactions and state of iron atoms in the $Bi_5Nb_{3-3x}Fe_{3x}O_{15-\delta}$ solid solutions, *Russ J Gen Chem*, **87**(3), 373–380.

50. Chezhina NV, Korolev DA, Fedorova AV, Zhuk NA, Filippova MV, Feltsinger LS, Lutoev VP, Makeev BA, Shevchuk SS, Nizovtsev AN (2017) Structure, magnetic, and electrical properties of bismuth niobates doped with *d*-elements: XVII Magnetic properties of the $Bi_5Nb_{3-3x}Mn_{3x}O_{15-\delta}$ manganese containing solid solutions, *Russ J Gen Chem*, **87**(10), 2251–2257.

51. Shannon RD, Prewitt CT (1969) Effective ionic radii in oxides and fluorides, *Acta Cryst B*, **25**, 925–946.

52. McCauley RA (1980) Structural characteristics of pyrochlore formation, *J Appl Phys*, **51**(1), 290–294.

53. Hoekstra HR, Siegel S (1968) Synthesis of thallium platinate at high pressure, *Inorg Chem*, **7**(1), 141–145.

54. Chakoumakos BC (1984) Systematics of the pyrochlore strucrure type, ideal $A_2B_2X_6Y$, *J Solid State Chem*, **53**(1), 120–129.

55. Vanderah TA, Levin I, Lufaso MW (2005) An unexpected crystal-chemical principle for the pyrochlore structure, *Eur J Inorg Chem*, **2005**(14), 2895–2901.

56. Torii Y, Hasegawa K (1977) The formation and structure of $Bi_{1.34}CrNbO_6$, *Bull Chem Soc Jpn*, **50**(10), 2638–2642.

57. Seshadri R (2006) Lone pairs in insulating pyrochlores: Ice rules and high-k behavior, *Solid State Sci*, **8**(3–4), 259–266.

58. Filoti GV, Rosenberg M, Kuncser V, Seling B, Fries T, Spies A, Kemmler-Sack S (1998) Magnetic properties and cation distribution in iron containing pyrochlores, *J Alloys Compd*, **268**(1–2), 16–21.

59. Valant M, Davies PK (2000) Crystal chemistry and dielectric properties of chemically substituted $(Bi_{1.5}Zn_{1.0}Nb_{1.5})O_7$ and $Bi_2(Zn_{2/3}Nb_{4/3})O_7$ pyrochlores, *J Am Ceram Soc*, **83**(1), 147–153.

60. Levin I, Amos TG, Nino JC, Vanderah TA, Randall CA, Lanagan MT (2002) Structural study of an unusual cubic pyrochlore $Bi_{1.5}Zn_{0.92}Nb_{1.5}O_{6.92}$, *J Solid State Chem*, **168**(1), 69–75.

61. Tan KB, Lee CK, Zainal Z, Miles GC, West AR (2005) Stoichiometry and doping mechanism of the cubic pyrochlore phase in the system Bi_2O_3–ZnO–Nb_2O_5, *J Mater Chem*, **15**, 3501–3506.

62. Sirotinkin VP, Bush AA (2003) Preparation and dielectric properties of $Bi_{1.5}MNb_{1.5}O_7$ (M = Cu, Mg, Mn, Ni, Zn) pyrochlore oxides, *Inorg Mater*, **39**(9), 974–977.

63. Tan KB, Lee CD, Zainal Z, Khaw CC, Tan YP, Shaari H (2008) Reaction study and phase formation in Bi_2O_3-ZnO-Nb_2O_5 ternary system, *Pacific J Sci Tech*, **9**(2), 468–479.

64. Liu Y, Withers RL, Chen H, Li Q, Tan H (2011) Raman spectra, photoluminescence and dielectric relaxation in $Bi_{1.5}ZnNb_{1.5}O_7$ pyrochlore, *Curr Appl Phys*, **11**(3), S171–S174.

65. Vanderah TA, Lufaso MW, Adler AU, Levin I, Nino JC, Provenzano V, Schenck PK (2006) Subsolidus phase equilibria and properties in the system Bi_2O_3:$Mn_2O_{3\pm x}$:Nb_2O_5, *J Solid State Chem*, **179**(12), 3467–3477.

66. Lufaso MW, Vanderah TA, Pazos IM, Levin I, Roth RS, Nino JC, Provenzano V, Schenck PK (2006) Phase formation, crystal chemistry,

and properties in the system Bi_2O_3–Fe_2O_3–Nb_2O_5, *J Solid State Chem*, **179**(12), 3900–3910.

67. Vanderah TA, Siegrist T, Lufaso MW, Yeager MC, Roth RS, Nino JC, Yates S (2006) Phase formation and properties in the system Bi_2O_3: $2CoO_{1+x}$:Nb_2O_5, *Eur J Inorg Chem*, **2006**(23), 4908–4914.
68. Nguyen HB, Norén L, Liu Y, Withers RL, Wei X, Elcombe MM (2007) The disordered structures and low temperature dielectric relaxation properties of two misplaced-displacive cubic pyrochlores found in the Bi_2O_3–$M^{II}O$–Nb_2O_5 (M = Mg, Ni) systems, *J Solid State Chem*, **180**(9), 2558–2565.
69. Valant M, Suvorov D (2005) The Bi_2O_3-Nb_2O_5-NiO phase diagram, *J Am Ceram Soc*, **88**(9), 2540–2543.
70. Hector AL, Wiggin SB (2004) Synthesis and structural study of stoichiometric $Bi_2Ti_2O_7$ pyrochlore, *J Solid State Chem*, **177**(1), 139–145.
71. Kennedy BJ (1997) Structural trends in Bi containing pyrochlores: The structure of $Bi_2Rh_2O_7$, *Mater Res Bull*, **32**(5), 479–483.
72. Yordanov SP, Ivanov I, Carapanov ChP (1998) Dielectric properties of the ferroelectric $Bi_2Ti_2O_7$ ceramics, *J Phys D: Appl Phys*, **31**, 800–806.
73. Roberto Esquivel-Elizondo J, Hinojosa BB, Nino JC (2011) $Bi_2Ti_2O_7$: It is not what you have read, *Chem Mater*, **23**(22), 4965–4974.
74. Rodriguez-Carvajal J (1993) Recent advances in magnetic structure determination by neutron powder diffraction, *Phys B: Cond Matter*, **192**, 55–60.
75. Koroleva MS, Piir IV, Ryabkov YuI, Korolev DA, Chezhina NV (2013) Synthesis and properties of chromium-containing bismuth titanates with the pyrochlore structure, *Russ Chem Bull*, **62**(2), 408–411.
76. Chezhina NV, Piir IV (2004) State of chromium in solid solutions of multicomponent bismuth niobates with pyrochlore structure, *Russ J Gen Chem*, **74**(7), 980–982.
77. Piir IV, Koroleva MS, Korolev DA, Chezhina NV, Semenov VG, Panchuk VV (2013) Bismuth iron titanate pyrochlores: Thermostability, structure, properties, *J Solid State Chem*, **204**, 245–250.
78. Piir IV, Sekushin NA, Grass VE, Ryabkov YI, Chezhina NV, Nekipelov SV, Sivkov VN, Vyalikh DV (2012) Bismuth manganese titanate: Crystal structure and properties, *Solid State Ion*, **225**, 464–470.

79. Gilbert B, Frazer BH, Belz A, Conrad PG, Nealson KH, Haskel D, Lang JC, Srajer G, De Stasio G (2003) Multiple scattering calculations of bonding and x-ray absorption spectroscopy of manganese oxides, *J Phys Chem A*, **107**, 2839–2847.

80. Efremov DV, Van den Brink J, Khomskii DI (2004) Bond-versus site-centred ordering and possible ferroelectricity in manganites, *Nat Mater*, **3**, 853–858.

81. Subramanian MA, Clearfield A, Umarji AM, Shenoy GK, Subba Rao GV (1984) Synthesis and solid state studies on manganese antimonate ($Mn_2Sb_2O_{7-x}$) and manganese cadmium antimonate $(Mn_{1-x}Cd_x)_2Sb_2O_{7-x}$ pyrochlore, *J Solid State Chem*, **52**(2), 124–129.

Index

For Product Safety Concerns and Information please contact our EU
representative GPSR@taylorandfrancis.com
Taylor & Francis Verlag GmbH, Kaufingerstraße 24, 80331 München, Germany

www.ingramcontent.com/pod-product-compliance
Ingram Content Group UK Ltd.
Pitfield, Milton Keynes, MK11 3LW, UK
UKHW022158190926
475702UK00003B/96

* 9 7 8 9 8 1 4 8 0 0 5 5 6 *